Plant
Systematics
and
Evolution   Supplementum 7

M. Hesse, E. Pacini, M. Willemse (eds.)

The Tapetum

Cytology, Function, Biochemistry and Evolution

Springer-Verlag Wien GmbH

Prof. Dr. M. HESSE
Institut für Botanik und Botanischer Garten
Universität Wien, Austria

Prof. Dr. E. PACINI
Dipartimento di Biologia Ambientale
Universitá degli Studi di Siena, Italy

Prof. Dr. M. WILLEMSE
Department of Plant Cytology and Morphology
Agricultural University Wageningen, The Netherlands

Typesetting: Thomson Press (India) Ltd., New Dehli, 110001

Printed on acid-free and chlorine-free bleached paper

With 98 Figures

ISSN 0172-6668
ISBN 978-3-211-82486-3   ISBN 978-3-7091-6661-1 (eBook)
DOI 10.1007/978-3-7091-6661-1

Preface

Dear Reader,

may we take the opportunity to inform you on the purpose and the background of this Volume.

As it is well known, the anther tapetum is a highly specialized, transient tissue nourishing (micro-)spores and pollen grains in *Bryophyta*, *Pteridophyta*, and *Spermatophyta*. Any tapetum malfunction causes pollen sterility; already this demonstrates clearly the importance of the tapetum for the reproductive life of all land plants. The exact knowledge of tapetum form, physiology and function is therefore indispensable not only in basic science, but also and especially in applied research, as e.g. in plant breeding and plant genetics.

Some seven years ago during M. H.'s stay at the University of Siena one of us (E.P.) came forward with the idea of a special tapetum conference. We felt that there was indeed urgent need for such a meeting: a tapetum symposium had never yet been organized; no special book been published; the existing reviews were only short and mostly out of date; and, for all, in recent years our insight in this field had grown enormously, basing on fascinating results using classical and new preparation or research techniques. At first we tried to organize a conference solely on tapetum and as soon as possible, but some invited specialists told us that they were "not yet ready" to present their results, and then the number of contributors was too small for an individual meeting. The idea of coupling the tapetum topic with a second related theme unfortunately could not be realized.

In 1990 during a palynology meeting in London and in accordance with our respected colleague Annick Le Thomas it was agreed that a special tapetum conference under the auspices of the VIII International Palynological Congress (IPC) should be organized. We invited our dear colleague and friend Michiel Willemse as third convener for this special conference because of his interest and competence. And indeed in Aix-en-Provence (France) on Friday, September 11th 1992, a Symposium was given, organized by M. Hesse, E. Pacini and M. T. M. Willemse, entitled "The tapetum: cytology, function, biochemistry and evolution".

Our original intention was to include contributions on the respective (male) tapetum tissues in mosses and ferns, in gymno- and angiosperms, and to focus on peculiar aspects of the (angiosperm) anther tapetum. Also closely related topics as e.g. the integumentary (female) tapetum were to be considered. Alas, in some topics we could not find adequate specialists because these themes had either been neglected for a long time or had never been investigated so far. So the contributions only cover some important aspects of gymno- and angiosperm tapetum tissues.

Nevertheless scientists from all over the world were invited and fourteen finally presented their results on current research topics (a dozen oral and two poster presentations). The themes ranged from developmental and cytological research based on conventional techniques up to specialized morphological, functional, or immunocytochemical research on significant tapetum functions, using advanced, sophisticated preparation methods. Most of these contributions are published as revised papers in the present volume. A review of present state-of-the-art preparation techniques in tapetum research by TEM, a survey of pertinent tapetum/pollen literature, and a Conspectus are added.

We are very grateful to our contributors, to the Head and Staff of VIII IPC Programme Committee for their effective co-operation, furthermore to our publisher, Springer-Verlag Wien New York, but particularly to Dr. I. KRISAI-GREILHUBER (the editorial secretary of Plant Systematics and Evolution) for her very welcome, time-consuming help in editing all the manuscripts.

We as the conveners of this tapetum symposium and simultaneously as the editors of this Volume wish that this book will help to stimulate and develop future tapetum research!

Vienna, April 1993   MICHAEL HESSE, ETTORE PACINI, MICHIEL WILLEMSE

Contents

Listed in Current Contents

Pl. Syst. Evol. [Suppl.] 7: 1–11 (1993)

Role of the tapetum in pollen and spore dispersal

E. Pacini and G. G. Franchi

Key words: *Embryophyta, Spermatophyta, Pteridophyta, Bryophyta.* – Anther tapetum, amoeboid tapetum, parietal tapetum, spore dispersal, pollination.

Abstract: Group-specific functions linked to the dispersal of pollen and spores are reported in all the groups of land plants. *Anthocerotopsida* tapetal cells envelop developing meiocytes and spores; after exine formation their walls thicken, the cytoplasm starts to degenerate, and elaters from more than one tapetal cell are formed. Elaters aid spore dispersal from the sporangium. In some *Marchantiopsida* the tapetal cells form elaters but each originates from one cell. In *Equisetophyta*, the spores have two or four elaters composed of lignin and connected with the exine. They are formed with the aid of the tapetal cytoplasm once exine is complete. The elaters of this group (called also adpressors) differ from the above in that they adhere to and originate connected with the exine. In *Polypodiophyta* like *Lecanopteris mirabilis*, the spores have superficial strands which facilitate adhesion to dispersing ants. One of the main features of *Spermatophyta* parietal tapetum is to produce orbicules. In strictly anemophilous species pollenkitt is not produced and pollen grains are dispersed individually. Since both exine and orbicules consist of sporopollenin, they have the same electrostatic charges. This is believed to aid expulsion from the anther at anthesis. Pollenkitt is a degeneration product of amoeboid and periplasmodial tapetum in *Magnoliophyta*. It forms a layer on pollen grains and has at least four different functions related to dispersal: (1) it causes the grains to unite in clumps and (2) to adhere to pollinator bodies; (3) it protects the pollen grain cytoplasm from sunlight; and (4) the oily and perfumed components attract pollinators.

The tapetum is a constant feature of *Embryophyta* anthers and sporangia. Pacini & al. (1985) distinguished eight types of tapeta on the basis of ontogeny and relationship to spores or developing pollen. They also proposed a hypothetical phylogenetic tree. Parietal tapetum is believed to be the oldest type from which all other types are derived from. The amoeboid tapetum of pteriodophytes and angiosperms is an example of evolutionary convergence. In the same paper the authors also list the specific functions of the tapetum in the various groups, aside from the common functions associated with pollen grain or spore nutrition and the formation of the exine. In subsequent papers (Pacini 1990, Pacini & Franchi 1991) the question was investigated to greater depth and 14 functions are currently recognized.

The tapetum produces nutrients, regulatory substances, and sporoderm precursors; the latter become polymerized in different sites and ways. Even the remnants of the degenerated tapetum have a function: they are involved in spore and pollen dispersal.

Land plant spores and pollen grains are small and their diameter commonly ranges from 10 to 100 μm. They are rarely bigger than 100 μm (ERDTMAN 1965). Dispersal of spores is commonly passive in cryptogams such as *Bryophyta* (RICHARDSON 1981) and *Pteridophyta* (PAGE 1979) but devices to facilitate spore dispersal start to appear in some mosses (*Anthocerotopsida* and *Marchantiopsida*) and are ± common in *Pteriodophyta*. Pollen dispersal in *Gymnospermae* is essentially passive viz. anemophilous. *Magnoliophyta* pollen may be distributed by wind, water or animals (Table 1).

Pteridophyta and *Bryophyta* spores are dispersed individually as monads. This is valid also for gymnosperm pollen grains, even in the rare entomophilous taxa (BINO & al. 1984). *Magnoliophyta* pollen grains may be dispersed as monads, as in strictly anemophilous species, or in groups formed in different ways (PACINI & FRANCHI 1991).

Devices aiding spore and pollen dispersal have different origins. Some, essentially of mechanical origin and sensitive to small changes in humidity, facilitate expulsion, as is the case with the moss peristome (PROCTOR 1984) and the catapult mechanism of some fern sporangia and anthers. Others are originating chemically, like the various substances which play a role in transfer by pollinators. The following paragraphs will elucidate dispersal mechanisms based on tapetal activity.

Bryophyta

Elaters are present in most *Anthocerotopsida* and *Marchantiopsida*. Elaters are dead tapetal cells with lignified walls and a narrow lumen; they are elongated and flexible (Table 1). They are intermingled with spores and react to changes of humidity thereby altering their configuration; their fast movements are responsible for spore ejection from the sporangium. At the beginning of sporangium development in *Anthocerotopsida* and *Marchantiopsida*, cells enveloping meiocytes and spores are acting as normal nutritive tapetal cells: at the end of this function their walls thicken and lignify and they die (PACINI & FRANCHI 1988). The elaters in the *Marchantiopsida* are single cells with spiral thickenings, while in the *Anthocerotopsida* no thickening occurs, and the sometimes even branched elaters consist of three to five cells in a row. All the tapetal cells of Anthocerotopsida and some of the *Marchantiopsida* degenerate, becoming elaters. A further difference between these two groups is that sporangium development is synchronous in *Marchantiopsida*. In the *Anthocerotopsida*, a meristem at the base of the cylindrical sporophyte is present, which is open at the top. All developmental stages occur within one sporangium simultaneously. This means that only a few spores are shed at a time. This case of individual reproductive structure with several stages present at the same time is unique among land plants.

Many species of *Splachnaceae*, a group of *Bryopsida* growing on carcasses and dung, have sticky spores dispersed "en masse" by flies and other coprophilic insects. These insects are attracted by often coloured and "attractively" odouring apophyses (PROCTOR 1984, RICHARSON 1981). No data are available on the nature of the adhesive substance of the spores but it may be pollenkitt from tapetum degeneration as in the *Magnoliophyta*.

Pteridophyta

In *Pteridophyta* sensu lato the tapetum is amoeboid, except in *Selaginella* (*Lycopodiophyta*), and is generally not involved in devices promoting spore dispersal. There are two important exceptions: the elaters of *Equisetophyta* and the "twisted cables of perispore" (sensu TRYON 1985) present in some myrmecophytic ferns like *Lecanopteris* (Table 1).

The *Equisetum* elaters, also called "adpressors", are quite different in structure and ontogenesis from those of the *Anthocerotopsida* and *Marchantiopsida*. Attached to the sporoderm, they are formed after its completion with the help of substances from the tapetum (UEHARA & KURITA 1989). They are small ribbons adhering to a narrow band of exine, sensitive to changes in humidity and actively moving the spore. *Bryophyta* elaters only facilitate spore release from the sporangium; in *Equisetophyta* the spores are generally dispersed from the sporangium by wind, and the adpressors mainly fulfill their function afterwards. They are activated every time a variation in environmental humidity occurs, even when the spore is dead. Larger distribution distances are achieved by this method than with the elaters of *Bryophyta*.

Table 1. Spore or pollen grain dispersal and tapetal devices involved in *Embryophyta*. [1]The spores of the coprophilic mosses, *Splachnaceae*, are dispersed by flies attracted by apophysal secretions (PROCTOR 1984). [2]Some ferns have viscin-like threads on the spore surface (TRYON 1985). [3]The nectary is situated next to the female part of some *Ephedra* spp. (BINO & al. 1984). [4]Something analogous to orbicules was described by LUGARDON (1981). + Present in almost all species, ± present in at least some members

Taxa	Dispersal			Tapetal devices						
	Passive, no devices	Active	Induced, by reward such as nectar and/or pollen	Elaters	Adpressors	Orbicules	Viscin threads	Elastoviscin	Pollenkitt	Tryphine (some *Cruciferae*)
Bryophyta										
Bryopsida	+		+[1]							
Marchantiopsida	+	+		±						
Anthocerotopsida		+		+						
Psilotophyta	+									
Lycopodiophyta	+									
Equisetophyta		+			+					
Polypodiophyta	+	+[2]				+[4]	±			
Gymnospermae	+		+[3]			+[3]				
Magnoliophyta										
Magnoliopsida	+	+	+			±	±		±	±
Liliopsida	+	+	+			±		±	±	

In myrmecophytic ferns, the twisted cables of perispore are very similar to angiosperm viscin threads. This device not only allows dispersal of all the spores of a sporangium in a cluster (WALKER 1985) but also keeps the spores as far as possible away from the ant bodies. Ants emit a volatile secretion which destroys molds growing in the nest and also kills pollen grins (BEATTIE & al. 1985) and spores.

Gymnospermae

Pollen dispersal in *Gymnospermae* is mainly anemophilous, and male cones are exposed to the wind. No devices, like viscin threads or pollenkitt, exist to promote pollen adhesion or its transport in clumps (BINO & al. 1984, HESSE 1984a).

Gymnosperm pollen has a clearly defined and therefore much more limited destination: The right landing place for its pollen (the female counterpart) is much more limited than that of *Bryophyta* and *Pteridophyta* spores (the ground). *Gymnospermae* have not developed different ways of conveying pollen grains towards the ovules; their only device are aerial sacci or wings, empty extensions of the exine. The first developmental stages of these sacci take place during the late tetrad stage when the primexine is formed. Their exine is completed with the help of the tapetum. They are inflated by gases, reaching their final size and shape during the early microspore stage (DICKINSON & BELL 1970, WILLEMSE 1971, KURMANN 1989). Sacci or wings, usually two per grain, cause the species-specific geometry of the pollen grains of many genera of the gymnosperms (ERDTMAN 1965). The shape of the pollen grain and of the female cone must be reasonably constant at pollination, because they are responsible that the pollen reaches the micropyle (NICKLAS 1985). Pollination in gymnosperms occurs by a micro-cloud of pollen grains which reaches the top of a female cone. There they are routed by the scales on different trajectories, and only the "right ones" reach the pollination drop at the top of the ovule (NICKLAS 1985).

Magnoliophyta

Pollen grains of gymnosperms are dispersed individually. In *Magnoliophyta*, however, anemophilous taxa shed single pollen grains, while in entomophilous taxa the transport of compound pollen occurs in several ways. This kind of transport is probably due to the fact that the megasporangium (the ovary) may contain from one to many thousands of female gametes, depending on the group, all to be fertilized therein. Pollen grain aggregation by clumping is realized in two ways, irrespective of the vector: (1) pollen grains remain together because they have joint walls, i.e. compound pollen s. str.; (2) single grains or tetrads are glued together. In both cases the tapetum is responsible for this mass transport by forming exine precursors or sticky material. Table 1 lists tapetum-dependent devices for pollen dispersal.

Ubisch bodies (orbicules) are present in *Gymnospermae* and *Magnoliophyta* with a secretory tapetum. In cryptogams they have only been reported in *Psilotum triquetrum* and several *Polypodiophyta* (LUGARDON 1981). They have never been found coexisting with devices such as viscin threads and elastoviscin. They are the only product of the degenerating tapetum in strictly anemophilous taxa such as *Gymnospermae*, *Urticaceae*, and *Poaceae*. In other, not so strict anemophilous taxa, as in *Salicaceae*, or in most entomophilous taxa they coexist with pollenkitt. Only few taxa with a secretory tapetum with a strong entomophilous syndrome, such as the pumkin (*Cucurbita pepo*) (CIAMPOLINI & al. 1993) and orchids (FITZGERALD & al. 1993) lack Ubisch bodies entirely (Table 2). Orbicules are species-specific and in some cases they have the same ornamentation as the exine (HESSE 1986).

A possible function of Ubisch bodies was first proposed by HESLOP-HARRISON & DICKINSON (1969): The Ubisch bodies might form a non-wettable surface on the locular wall, effecting an easy detachment and dispersal of pollen. This was confirmed by ECHLIN (1971) and KEIJZER (1987). In entomophilous species the pollen grains usually stay in the anther until pollinators actively or passively collect them. In other – entomophilous – taxa with orbicules the pollen should not stay long in the loculus after anther opening. Since exine and orbicules both consist of sporopollenin their surfaces are electrically charged in the same way; this leads to mutual repulsion and enhances pollen expulsion. So, especially in anemophilous taxa, when the anthers open, the pollen grains are expelled. This hypothesis is supported by the fact that orbicules are also present in some *Pteriodophyta* (LUGARDON 1981). This is one of the main differences in anther features between anemophilous and entomophilous species. Further on, the simultaneous occurrence of pollenkitt and orbicules may be related to a lack of specialization in the pollination syndromes, e.g., in *Apiaceae, Oleaceae, Rosaceae*, and *Rutaceae*.

Viscin threads are thin ropes of sporopollenin continuous with the outer exine layer (HESSE 1984b). They are found in some dicot families dispersing pollen grains in tetrads, e.g., *Onagraceae* p.p. and *Ericaceae* p.p., or monads, e.g. *Epilobium* (*Onagraceae*) or *Leguminosae: Caesalpinioideae* (PATEL & al. 1985, SKVARLA & al. 1975, WAHA 1984). In either case viscin threads are associated with mass transport and/or adhesion to an insect body (HESSE 1980a). HESSE (1981) presents the physicochemical properties of pollenkitt and viscin threads, two analogue devices with the same function, and compares these two modes of pollen transport.

Elastoviscin occurs only in orchids with tetrads and monads. This viscous substance is formed by the tapetal cytoplasm as unsaturated lipid droplets from the microspore stage onwards. Before anthesis the droplets fuse and flow into the loculus, glueing the pollen grains together (SCHILL & WOLTER 1985, YEUNG & LAW 1987, WOLTER & al. 1988). SCHILL & WOLTER (1986), comparing the ontogeny and properties of elastoviscin and pollenkitt, find that the main differences consist in the stickiness (elastoviscin is much more viscous) and the number of pollen grains (much higher in orchids). A further difference: in many groups of orchids elastoviscin is not produced by all the tapetal cells, but only by a restricted and well localized group of them (SCHILL & WOLTER 1986, HESSE & BURNS-BALOGH 1984).

Pollenkitt is the final product of the degeneration of the secretory and amoeboid tapetum (PACINI & KEIJZER 1989). It covers the pollen surface just before the opening of the anther, consisting only of hydrophobic material derived from blends of elaioplast and cytoplasmic lipids. Sometimes pollenkitt is found together with orbicules (Table 2).

Several functions can be ascribed to pollenkitt (Table 3) according to the pollination strategy and pollen grain size. Some functions are due to the viscous nature of pollenkitt, namely to stick pollen grains together in the anther until the arrival of pollinators and during dispersal. Other properties depend on the chemical

Table 2. List of angiosperm species with secretory tapetum reported to have orbicules (O), pollenkitt (P) or tryphine (T). *These authors refer to pollenkitt as "tryphine"

Taxa	Devices OPT	Reference
Magnoliopsida		
Apiaceae		
Smyrnium perfoliatum L.	OP	Weber (1991)
Austrobaileyaceae		
Austrobaileya maculata L.T. White	OP	Zavada (1984)
Caryophyllaceae		
Silene dioica (L.) Clairv.	OP	Audran & Batcho (1981)*
Cruciferae		
Brassica oleracea L.	T	Murgia & al. (1991)
Raphanus sativus L.	T	Dickinson & Lewis (1973)
Cucurbitaceae		
Cucurbita pepo L.	P	Ciampolini & al. (1993)
Ericaceae		
Andromeda japonica Thunb.	OP	Hesse (1979a)
Calluna vulgaris H. Hull	OP	Hesse (1979a)
Erica arborea L.	OP	Hesse (1979a)
Euphorbiaceae		
Euphorbia palustris L.	OP	Hesse (1986)
Hamamelidaceae		
Hamamelis virginiana L.	OP	Hesse (1981)
Mimosaceae		
Acacia spec.	O	Kenrick & Knox (1979)
Oleaceae		
Olea europaea L.	OP	Pacini & Juniper (1979)
Polygalaceae		
Securidaca longepedunculata Fresen.	OP	Coetzee & Robbertse (1985)
Polygonaceae		
Fagopyrum esculentum Moench.	P	Hesse (1979b)
Polygonum bistorta L.	P	Hesse (1979b)
Rheum rhabarbarum L.	P	Hesse (1979b)
Rosaceae		
Prunus avium L.	OP	Pacini & al. (1986)
Rutaceae		
Citrus limon Burm.	OP	(our unpublished data)
Salicaceae		
Populus nigra L.	OP	Hesse (1979a)
Salix caprea L.	OP	Hesse (1979a)
Solanaceae		
Lycopersicum peruvianum Mill.	O	Pacini & Juniper (1984)
Tiliaceae		
Tilia platyphyllos Scop.	OP	Hesse (1979a)
T. tomentosa Moench.	OP	Hesse (1979a)
Urticaceae		
Parietaria judaica L.	O	(our unpublished data)

Table 2. (*Continued*)

Taxa	Devices OPT	Reference
P. officinalis L.	O	FRANCHI & PACINI (1981)
Urtica dubia FORSKAL	O	FRANCHI & PACINI (1981)
Liliopsida		
Cyperaceae		
Carex baldensis L.	OP	HESSE (1980b)
Heleocharis palustris L. ROEMER & SCHULTZ	OP	CARNIEL (1971)
Liliaceae		
Lilium cv. Enchantement	OP	REZNICKOVA & WILLEMSE (1980)
Trillium kamtschaticum LEDEB.	OP	TAKAHASHI (1987)
Poaceae		
Lolium perenne L.	O	PACINI & al. (1992)
Strelitziaceae		
Strelitzia reginae AIT.	OP	KRONESTEDT-ROBARDS & ROWLEY (1989)

Table 3. Functions of pollenkitt

Primary function	Secondary function
1 To hold pollen grains together	– in anther until collection by insects – in anther and in air. Physical – chemical properties of pollenkitt, and pollen grain size determine number of grains per group. Deposition of whole group on stigma surface; this occurs in both ento-and anemophilous species. – in air, sticking them to the insect during flight.
2 To protect against effects of solar radiation, i.e.	– damage to vegetative cell cytoplasm. – mutations in generative and sperm cells.
3 To protect pollen from further water loss	
4 To determine pollen colour.	
5 To maintain sporophytic proteins inside exine cavities	
6 To attract potential insect pollinators.	

nature of pollenkitt, namely protection against solar radiation and water loss during exposure and flight. Others are due to colour and smell which attract pollinators. In *Lagerstroemia indica* which has two sets and kinds of anthers, the inner one producing nutritive pollen to feed pollinators and the outer one producing real, genuine pollen, the pollenkitt differs in colour (Pacini & Bellani 1986) and physicochemical properties (Fitzgerald & al. 1993).

Tryphine is not as common as pollenkitt, being described only in some *Cruciferae*. It consists of a mixture of hydrophilic and hydrophobic material derived from the degeneration of intact organelles of tapetal cells and is spread over the microspore surface after the rupture of tapetal plasmamembrane (Dickinson & Lewis 1973, Pacini & Franchi 1991). Its properties and functions are probably the same as pollenkitt.

The flexuous pollen grains of marine monocots are coated with a substance, probably derived from tapetal degeneration (Ducker & Knox 1976). This glue and the thread-like shape of the grains help to stick the pollen grains together when the anthers open, allowing a gradual dispersion with tidal movements (Cox & Knox 1989).

Conclusions

In compound pollen (i.e. tetrads, polyads, massulae, and probably also pollinia) the number of grains in the dispersal unit is constant and always a multiple of four. This number of grains per unit is equal to the male gametes brought together on a single stigma, and in the case of pollinia, all the pollen grains of a flower land on the same stigma.

When the pollen connecting agents are viscin threads, elastoviscin, pollenkitt or tryphine, the number of grains constituting the dispersal unit is very variable, but the number per cluster is never as large as in multiple pollen grains (polyads). This number depends on: (a) the size of the grain and consequently its weight – the larger the grain the fewer the number; (b) the amount of the connecting agent and its distribution on the surface of each pollen grain and within the pollen grain aggregate; (c) the physicochemical properties of the connecting agent; and (d) the area and nature of the surface of the pollinator where grains can adhere to.

In conclusion, the tapetum has essentially two functions in dispersal. The first is to contribute to spore and pollen grain expulsion by means of elaters, adpressors, and perhaps orbicules. This function seems to be a primitive feature because it is present in early land plants (i.e. elaters and adpressors) and in *Gymnospermae* (i.e. orbicules) but since in *Magnoliophyta* anemophily can not be considered a primitive feature (Crepet & Friis 1992), the case remains unsettled so far.

The second – evident – function, exemplified in *Magnoliophyta*, is to aggregate pollen grains for mass transport, and to stick these pollen masses to the pollinator's body. This second – well proven – function increases the pollinator's efficiency and is of importance in phylogenetics and ecology.

Research performed under CNR program "Reproductive Ecology."

References

Audran, J. C., Batcho, M., 1981: Cytochemical and intrastructural aspects of pollen and tapetum ontogeny in *Silene dioica* (*Caryophyllaceae*). – Grana **20**: 65–80.

BEATTIE, A. J., TURNBULL, C., HOUGH, T., JOBSON, S., KNOX, R. B., 1985: The vulnerability of pollen and fungal spores to ant secretions: evidence and some evolutionary implications. – Amer. J. Bot. **72**: 606–614.

BINO, R. J., DEVENTE, N., MEEUSE, A. D. N., 1984: Entomophily in the dioecious gymnosperm *Ephedra aphylla* FORSK. (– *E. alte* C. A. MEY.) with some notes on *E. campylopoda* C. A. MEY. – Proc. Kon. Nederl. Akad. **87**: 15–24.

CARNIEL, K., 1971: Über die lamelläre Struktur und die Herkunft des Pollenkittes bei *Heleocharis palustris*. – Österr. Bot. Z. **119**: 464–474.

CIAMPOLINI, F., NEPI, M., PACINI, E., 1993: Tapetum development in *Cucurbita pepo*. – Pl. Syst. Evol. [Suppl.] **7**: 13–22.

COETZEE, H., ROBBERTSE, P. J., 1985: Pollen and tapetal development in *Securidaca longepeduculata*. – S. African J. Bot. **51**: 111–124.

COX, P., KNOX, B. R., 1989: Two-dimensional pollination in hydrophilous plants. – Amer. J. Bot. **76**: 164–175.

CREPET, W. L., FRIIS, E. M., 1992: The evolution of insect pollination in angiosperms. – In FRIIS, E. M., CHALONER, W. G., CRANE, P. R., (Eds): The origin of angiosperms and their biological consequences, pp. 181–201. – Cambridge: Cambridge University Press.

DICKINSON, H. G., BELL, P. R., 1970: The development of the sacci during pollen formation in *Pinus banksiana*. – Grana **10**: 101–118.

LEWIS, D., 1973: The formation of tryphine coating the pollen grains of *Raphanus* and its properties relating to self-incompatibility system. – Proc. Roy. Soc. London B **184**: 149–165.

DUCKER, S., KNOX, R. B., 1976: Submarine pollination in seagrasses. – Nature **263**: 705–706.

ECHLIN, P., 1971: Production of sporopollenin by the tapetum. – In MUIR, P. M., VAN GIJZEL, P., SHAW, G., (Eds): Sporopollenin, pp. 220–241. – London: Academic Press.

ERDTMAN, G., 1965: Pollen and spore morphology – Plant taxonomy – *Gymnospermae, Bryophyta* (text). An introduction to palynology. III. – Stockholm: Almqvist & Wiksell.

FITZGERALD, M. A., KNOX, R. B., CALDER, D. M., PACINI, E., 1993: To glue or not to glue. – Ann. Bot. (in press).

FRANCHI, G. G., PACINI, E., 1981: Gli orbicoli: formazione, morfologia e loro significato. – Giorn. Bot. Ital. **115**: 133.

HESLOP-HARRISON, J., DICKINSON, H. G., 1969: Time relationships of sporopollenin synthesis associated with tapetum and microspores in *Lilium*. – Planta **84**: 199–214.

HESSE, M., 1979a: Entwicklungsgeschichte und Ultrastruktur von Pollenkitt und Exine bei nahe verwandten entomo- und anemophilen Angiospermen: *Salicaceae, Tiliaceae* und *Ericaceae*. – Flora **168**: 540–557.

– 1979b: Entwicklungsgeschichte und Ultrastruktur von Pollenkitt und Exine bei nahe verwandten entomo- und anemophilen Angiospermen: *Polygonaceae*. – Flora **168**: 558–577.

– 1980a: Zur Frage der Anheftung des Pollens an blütenbesuchende Insekten mittels Pollenkitt und Viscinfäden. – Pl. Syst. Evol. **133**: 135–148.

– 1980b: Entwicklungsgeschichte und Ultrastruktur von Pollenkitt und Exine bei nahe verwandten entomophilen und anemophilen Angiospermensippen der *Alismataceae, Liliaceae, Juncaceae, Cyperaceae, Poaceae* und *Araceae*. – Pl. Syst. Evol. **134**: 229–267.

– 1981: Pollenkitt and viscin threads: their role in cementing pollen grains. – Grana **20**: 145–152.

– 1984a: Pollenkitt is lacking in *Gnetatae: Ephedra* and *Welwitschia*; further proof for its restriction to the angiosperms. – Pl. Syst. Evol. **144**: 9–16.

– 1984b: An exine architecture model for viscin threads. – Grana **23**: 69–75.

– 1986: Orbicules and the ektexine are homologous sporopollenin concretions in *Spermatophyta*. – Pl. Syst. Evol. **153**: 37–48.

– Burns-Balogh, P., 1984: Pollen and pollinarium morphology of *Habenaria* (*Orchidaceae*). – Pollen & Spores **26**: 385–400.

Kenrick, J., Knox, R. B., 1979: Pollen development and cytochemistry in some Australian species of *Acacia*. – Austral. J. Bot. **27**: 413–427.

Keijzer, C. J., 1987: The processes of anther dehiscence and pollen dispersal. II. – New Phytol. **105**: 499–509.

Kronestedt-Robards, E. C., Rowley, J. R., 1989: Pollen grain development and tapetal changes in *Strelitzia reginae* (*Strelitziaceae*). – Amer. J. Bot. **76**: 856–870.

Kurmann, M. H., 1989: Pollen wall formation in *Abies concolor* and a discussion on wall layer homologies. – Canad. J. Bot. **67**: 2489–2504.

Lugardon, B., 1981: Les globules des Filicinées, homologues des corps d'Ubisch des Spermaphytes. – Pollen & Spores **23**: 93–124.

Murgia, M., Charzynska, M., Rougier, M., Cresti, M., 1991: Secretory tapetum of *Brassica oleracea* L.: polarity and ultrastructural features. – Sex. Pl. Reprod. **4**: 28–35.

Nicklas, K. L., 1985: The aerodynamics of wind pollination. – Bot. Rev. **51**: 328–386.

Pacini, E., 1990: Tapetum and microspore function. – In Blackmore, S., Knox, R. B., (Eds): Microspores–evolution and ontogeny, pp. 213–237. – London: Academic Press.

– Bellani, L. M., 1986: *Lagerstroemia indica* L. pollen: form and function. – In Blackmore, S., Ferguson, I. K., (Eds): Pollen and spores: form and function, pp. 347–357. – London: Academic Press.

– Franchi, G. G., 1988: Sporophyte development in *Phaecoceros laevis* (L.) Prosk. – Giorn. Bot. Ital. **122** [Suppl. 1]: 93.

– – 1991: Diversification and evolution of the tapetum. – In Blackmore, S., Barnes, S. H., (Eds): Pollen and spores – patterns of diversification, pp. 301–316. – Oxford: Systematic Association, Clarendon Press.

– Juniper, P. E., 1979: The ultrastructure of pollen grain development in the olive (*Olea europaea*). II. Secretion by the tapetal cells. – New Phytol. **83**: 165–174.

– – 1984: The ultrastructure of pollen grain development in *Lycopersicum peruvianum*. – Caryologia **37**: 21–50.

– Keijzer, C. J., 1989: Ontogeny of intruding non-periplasmodial tapetum in the wild chicory, *Cichorium intybus* (*Compositae*). – Pl. Syst. Evol. **167**: 149–164.

– Bellani, L. M., Lozzi, R., 1986: Pollen, tapetum and anther development in two cultivars of sweet cherry (*Prunus avium*). – Phytomorphology **36**: 197–210.

– Franchi, G. G., Hesse, M., 1985: The tapetum: its form, function and possible phylogeny in *Embryophyta*. – Pl. Syst. Evol. **149**· 155–185.

– Taylor, P. E., Singh, M. B., Knox, R. B., 1992: Development of plastids in pollen and tapetum of rye-grass, *Lolium perenne* L. – Ann. Bot. **70**: 179–188.

Page, C. N., 1979: Experimental aspects of ferns ecology. – In Dyer, A. F., (Ed.): The experimental biology of ferns, pp. 552–589. – London: Academic Press.

Patel, V., Skvarla, J. J., Ferguson, I. K., Graham, A., Raven, P. H., 1985: The nature of threadlike structures and other morphological characters in *Jacqueshuberia* pollen (*Lequminosae: Caesalpinioideae*). – Amer. J. Bot. **72**: 407–413.

Proctor, M. C. F., 1984: Structure and ecological adaptation. – In Dyer, A. F., Duckett, J. G., (Eds): The experimental biology of Bryophytes, pp. 9–37. – London: Academic Press.

Reznickova, S. A., Willemse, M. T. M., 1980: Formation of pollen in the anther of *Lilium*: the function of the surrounding tissues in the formation of pollen and pollen wall. – Acta Bot. Neerl. **29**: 141–156.

Richardson, D. H. S., 1981: The biology of mosses. – London: Blackwell.

Schill, R., Wolter, M., 1985: Ontogeny of elastoviscin in the *Orchidaceae*. – Nordic J. Bot. **5**: 575–580.

– – 1986: On the presence of elastoviscin in all subfamilies of the *Orchidaceae* and the homology to pollenkitt. – Nordic J. Bot. **6**: 321–324.

SKVARLA, J. J., RAVEN, P. H., PRAGLOWSKI, J., 1975: The evolution of pollen tetrads. – Amer. J. Bot. **62**: 6–35.

TAKAHASHI, M., 1987: Development of omniaperturate pollen in *Trillium kamtschaticum* *(Liliaceae)*. – Amer. J. Bot. **74**: 1842–1852.

TRYON, A. F., 1985: Spores of myrmecophytic ferns. – Proc. Roy. Soc. Edinburgh **86B**: 105–110.

UEHARA, K., KURITA, S., 1989: An ultrastructural study of spore wall morphogenesis in *Equisetum arvense*. – Amer. J. Bot. **76**: 939–951.

WAHA, M., 1984: Zur Ultrastruktur und Funktion pollenverbindender Fäden bei *Ericaceae* und anderen Angiospermenfamilien. – Pl. Syst. Evol. **147**: 189–203.

WALKER, T. G., 1985: Spore filaments in the ant-fern *Lecanopteris mirabilis* – an alternative viewpoint. – Proc. Roy. Soc. Edinburgh **86B**: 111–114.

WEBER, M., 1991: The transfer of pollenkitt in *Smyrnium perfoliatum* (*Apiaceae*). – Ann. Bot. **68**: 63–68.

WILLEMSE, M. T. M., 1971: Morphological and fluorescence microscopical investigation on sporopollenin formation at *Pinus sylvestris* and *Gasteria verrucosa*. – In BROOKS, J., GRANT, P. R., MUIR, M., VAN GIJZEL, P., SHAW, G., (Eds): Sporopollenin, pp. 68–107. – London: Academic Press.

WOLTER, M., SEUFFERT, C., SCHILL, R., 1988: The ontogeny of pollinia and elastoviscin in the anthers of *Doritis pulcherrima* (*Orchidaceae*). – Nordic J. Bot. **8**: 77–88.

YEUNG, E. C., LAW, S. K., 1987: The formation of hyaline caudicle in two vandoid orchids. – Canad. J. Bot. **65**: 1459–1464.

ZAVADA, M. S., 1984: Pollen wall development of *Austrobaileya maculata*. – Bot. Gaz. **145**: 11–21.

Addresses of the authors: Prof. Dr E. PACINI, Department of Environmental Biology, Botanical Section, University of Siena, Italy. – Dr G. G. FRANCHI, Department of Pharmaceutical Sciences, University of Bologna, Bologna, Italy.

Pl. Syst. Evol. [Suppl.] 7: 13–22 (1993)

Tapetum development in *Cucurbita pepo* (*Cucurbitaceae*)*

F. Ciampolini, M. Nepi, and E. Pacini

Key words: *Cucurbitaceae, Cucurbita pepo.* – Anther tapetum, pollen, pollenkitt, sporophytic proteins.

Abstract: The tapetal cell walls are of uniform thickness before prophase; from leptotene onwards, those of the inner tangential and the proximal half of the radial wall swell. At late tetrad stage the swollen walls disappear, only the outer (distal) tangential and the distal part of the contiguous radial walls persist until degeneration. In prophase, plastids are undifferentiated, resembling those of microspore mother cells, but their development pattern changes from the first meiotic division: they divide actively in the early tetrad stage and differentiate into elaioplasts in the late microspore stage. Elaioplast and spherosome lipids contribute to pollenkitt formation. Degeneration of the tapetal cells results in a single lipid sphere for each cell; these spheres are pollenkitt precursors and cover the pollen grain in the late bicellular stage. Sporophytic proteins are deposited under the operculum during late microspore stage, face to face with gametophytic proteins localized in the pectocellulosic part of the pore. The tapetal cells degenerate during early bicellular pollen stage, in contrast to other species in which the degeneration takes place in the late microspore stage.

Tapetal cells delimit the loculus in many types of sporangia and anthers with secretory (parietal) tapetum. The loculus is only absent in a few cases: (1) in species with amoeboid (periplasmodial) tapetum; (2) in many *Compositae* with initially parietal tapetum, where the loculus later obliterates by the intrusion of amoeboid tapetal cells; (3) in anthers with compound pollen as in *Acacia* (Kenrick & Knox 1979) and in *Orchidaceae* with massulae or pollinia (Pacini & Franchi 1991).

All tapetal types of angiosperms can degenerate in the same way, leaving residues (Pacini 1990a). In all entomophilous taxa, the tapetum degenerates leaving pollenkitt which has several functions (Pacini & Franchi 1993), e.g., to unite pollen grains in clumps and to facilitate pollen adhesion to the pollinator.

The entomophilous species *Cucurbita pepo* is monoecious with unisexual flowers. Both flowers have nectaries and nectar is the only reward for pollinators. The nectaries are strongly dimorphic with regard to their position, their surface area, and the amount and composition of nectar (Nepi, unpubl. data). Pollen grains are large (200 µm in diameter) with 12 operculate pores.

The aim of the present paper is to describe cytological events of the secretory tapetum in relation to the developing pollen grain.

---

*This work was supported by a grant of M.U.R.S.T. 40%.

Material and methods

Single male flowers of *Cucurbita pepo* L. (*Cucurbitaceae*) in different developmental stages, were collected during the summers of 1990 and 1991 from plants cultivated in the Siena Botanical Gardens. Small portions of anthers were carefully dissected and fixed in 3% glutaraldehyde in 0.06 M cacodylate buffer at pH 7.2 for 2 h, rinsed in the same buffer, and postfixed in 1% osmium tetroxide for 1 h. After dehydration in an ethanol series, the samples were embedded in Spurr's low viscosity resin, cut on an LKB ultrotome III, stained with uranyl acetate and lead citrate, and observed with a JEOL JEM 100 at 80 KV.

For interference contrast microscopy semithin sections (2 μm) were cut from the same blocks as used for the ultrathin sections, and observed in water under immersion oil.

Flowers dissected as above were fixed in 4% freshly prepared paraformaldehyde in phosphate buffer at pH 7.2, dehydrated in an ethanol series, and embedded in LR white (London Resin Co. Ltd.). 2 μm semithin sections were tested for: (a) total polysaccharides with the PAS reaction preceded by aldehyde blockade with dimedone (O'BRIEN & MCCULLY 1981); (b) total proteins with bromophenol blue (PEARSE 1968); (c) callose with calcofluor white M2R (HUGHES & MCCULLY 1975).

Results

Twelve stages of tapetum-pollen development are recognized, named according to the developmental stages of the pollen grain (Fig. 1).

Late prophase. The callose wall is formed around the meiocytes (Fig. 1a). Tapetal cell walls are thin with the exception of the inner (proximal) tangential ones. Plasmodesmata occur only in the radial walls; plastids are undifferentiated.

Telophase of the second meiotic division. The callosic wall is more uniform (Fig. 1b). The thickness of the inner tangential wall of the tapetum has increased; the proximal part of the radial walls is swollen and plasmodesmata are still present. Small vacuoles surround the nucleus. Tapetal cells remain uninucleate until degeneration. Plastids are undifferentiated but divide because dumb-bells shaped plastids are often observed (Fig. 2).

Meiotic cleavage. The tapetal cells are as in the previous stage. The callose wall of the young microspores is formed centripetally (Fig. 1c).

Early tetrad stage. The thickness of the inner tangential tapetal cell wall decreases. (Fig. 1d). The callose wall of the young microspores is completed.

Middle tetrad stage. In the tetrads the sites of future pores are visible as plasma-membrane retraction. Primexine starts to form and the conical ornamentations of exine appear in the poral and interporal areas (Fig. 1e). The callose wall looses its compactness becoming spongy on the outside (Fig. 3). Small vacuoles of tapetal cytoplasm fuse and plastids continue to divide (Fig. 3). The unswollen walls of tapetal cells are unchanged but the swollen ones become multilayered with electron-opaque inclusions (Fig. 3). Active dictyosomes, producing numerous vesicles are found parallel to the plasmamembrane near dilated cisternae of endoplasmatic reticulum.

Late tetrad stage. The primexine of the microspores is divided into an electron-translucent layer and another one with the cylindrical ornamentations of the exine (Fig. 4). The inner tangential and the radial walls of the tapetal cells have completely disappeared (Fig. 1f); only some faint granular electron opaque debris and strong

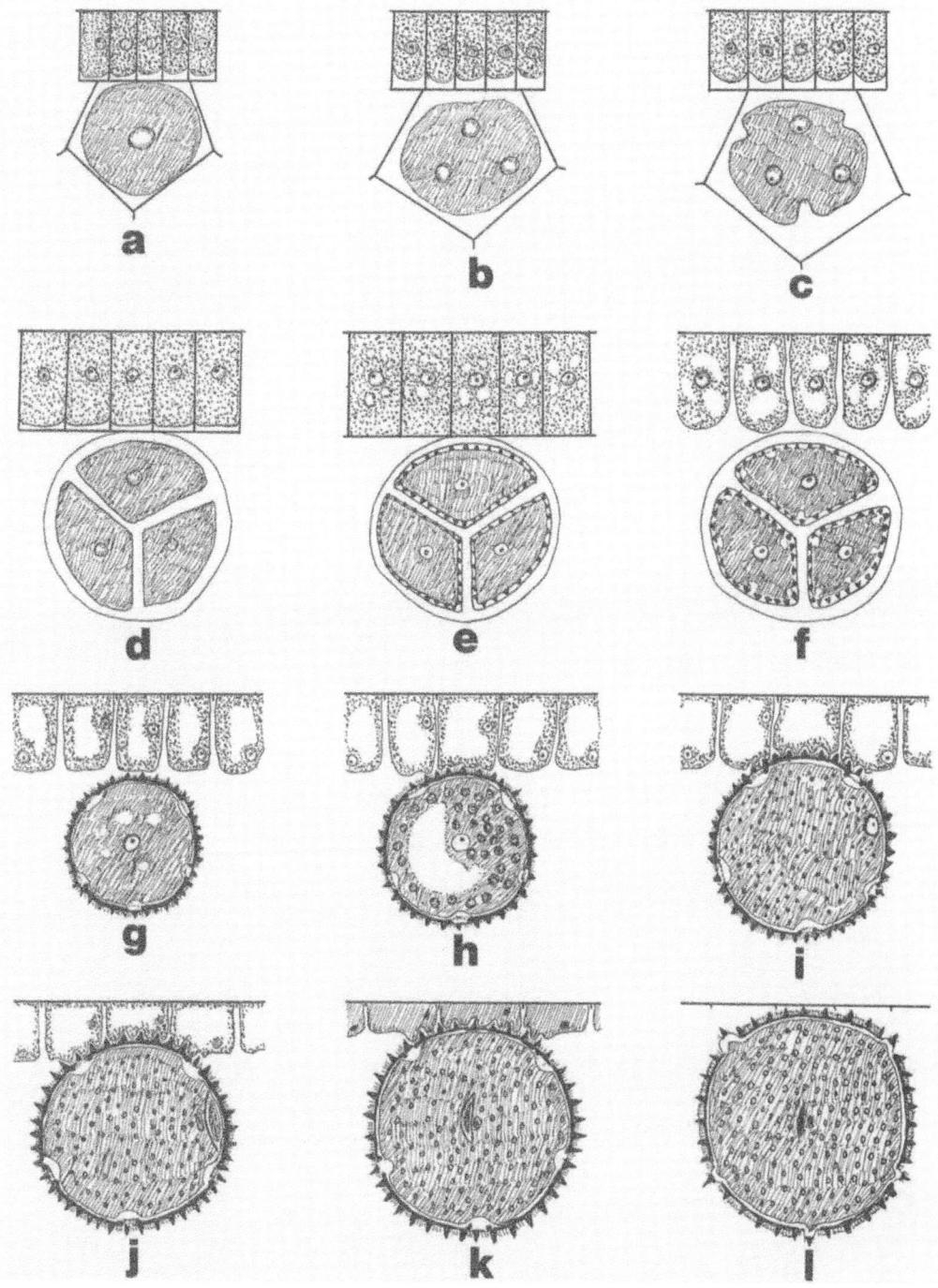

Fig. 1. Tapetum and pollen development in *Cucurbita pepo*, schematically. Twelve stages: *a* late prophase; *b* second meiotic division; *c* meiotic cleavage; *d* early tetrad stage; *e* middle tetrad stage; *f* late tetrad stage; *g* early microspore stage; *h* middle microspore stage; *i* late microspore stage; *j* early bicelluar stage; *k* middle bicellular stage, and *l* late bicellular stage. The schematic drawings are in details (e.g. plastids in the microspores) not in scale, and the cytoplasmic contents of the microspores/pollen grains are only cursorily drawn

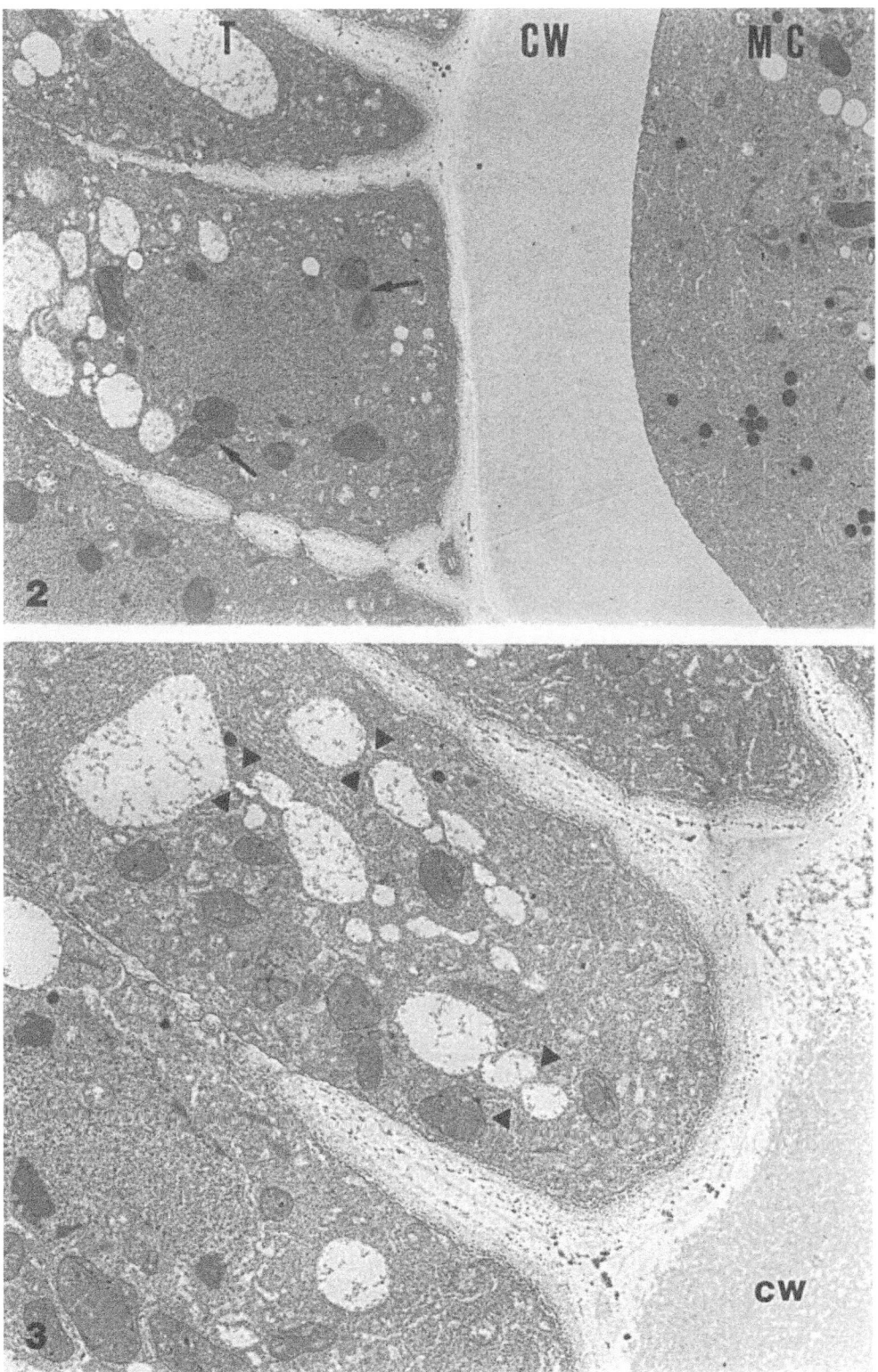

Figs. 2, 3. *Cucurbita pepo.* – Fig. 2. Tapetum (*T*) and meiocytes (*MC*) at telophase of the second meiotic division. The callose wall (*CW*) separates the tapetum from the meiocyte cytoplasm. The tapetal cell wall is thickened on the inner tangential wall and in the proximal part of the radial wall; plastids are dividing (arrows). × 5000. – Fig. 3. Tapetal cells and part of the spongy callose wall (*CW*) in the middle tetrad stage. The outer tangential and the distal part of the radial wall of the tapetal cell are thin, whereas the inner tangential and the proximal part of the radial wall are thickened and multilayered. Small vacuoles start to fuse (arrowheads). × 9000

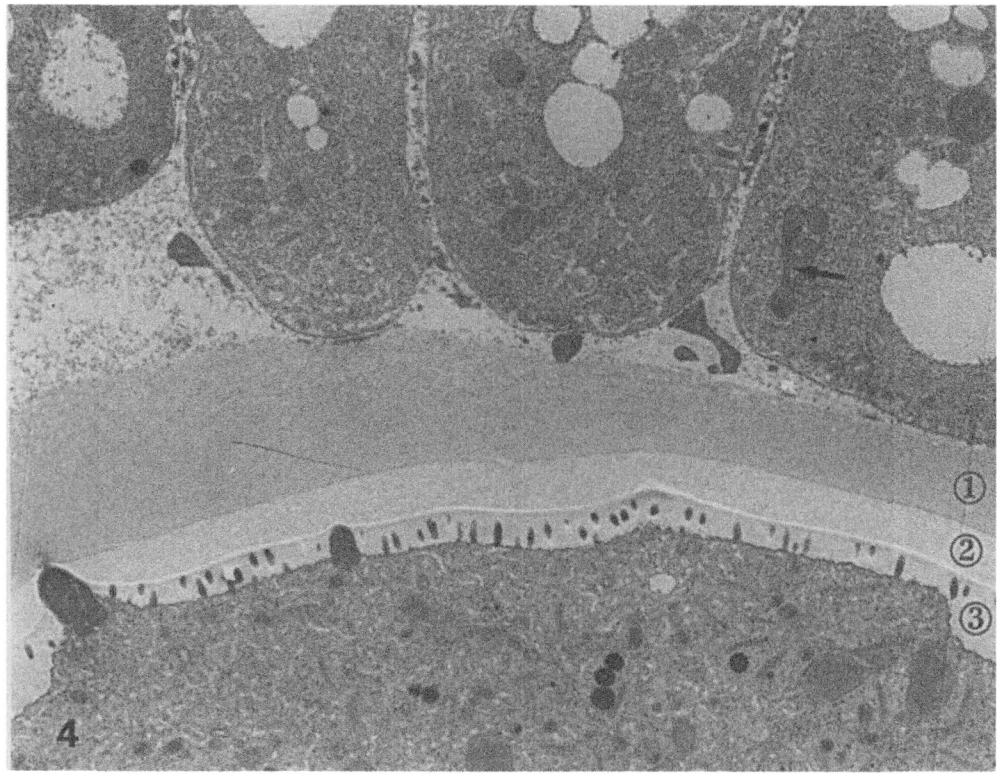

Fig. 4. *Cucurbita pepo*. Tapetal cells and a part of a microspore in the late tetrad stage. The medium dense callose wall 1 surrounds the tetrads. The microspore wall is divided in an outer electron translucent layer 2 and an inner layer 3 with large conical and smaller cylindrical spines. The tapetal cell walls degenerated, their residuals are granular deposits in the loculus. Irregular electron opaque bodies are visible in the loculus. Plastids continue to divide (arrow). × 8500

electron opaque deposits persist until later stages. The tapetal plastids continue to divide (Fig. 4).

Early microspores stage. The microspores have separated. The callose wall disappears. The exine is completed during this stage. Small vacuoles are found in the microspore (Fig. 1g). The tapetal cells reach their maximum size; their organelles are unchanged but each cell contains one big vacuole (Fig. 1g).

Middle microspore stage. The poral part of the exine is completed; poral and interporal intine starts to be formed. The microspores contain only one vacuole (Fig. 1h). The vacuolization continues by means of endoplasmic reticulum which forms small cytoplasm vesicles devoid of organelles which are enclosed by the big vacuole. Some of these vesicles of cytoplasm seem to be discharged into the tapetal intercellular spaces (Fig. 5) and the loculus.

Late microspore stage. Poral and interporal intine are completed and an electron opaque layer appears under the operculum (Fig. 7). This layer reacts positively to total protein tests such as bromophenol blue and consists of sporophytic proteins. The spongy exine of the operculum is different from the

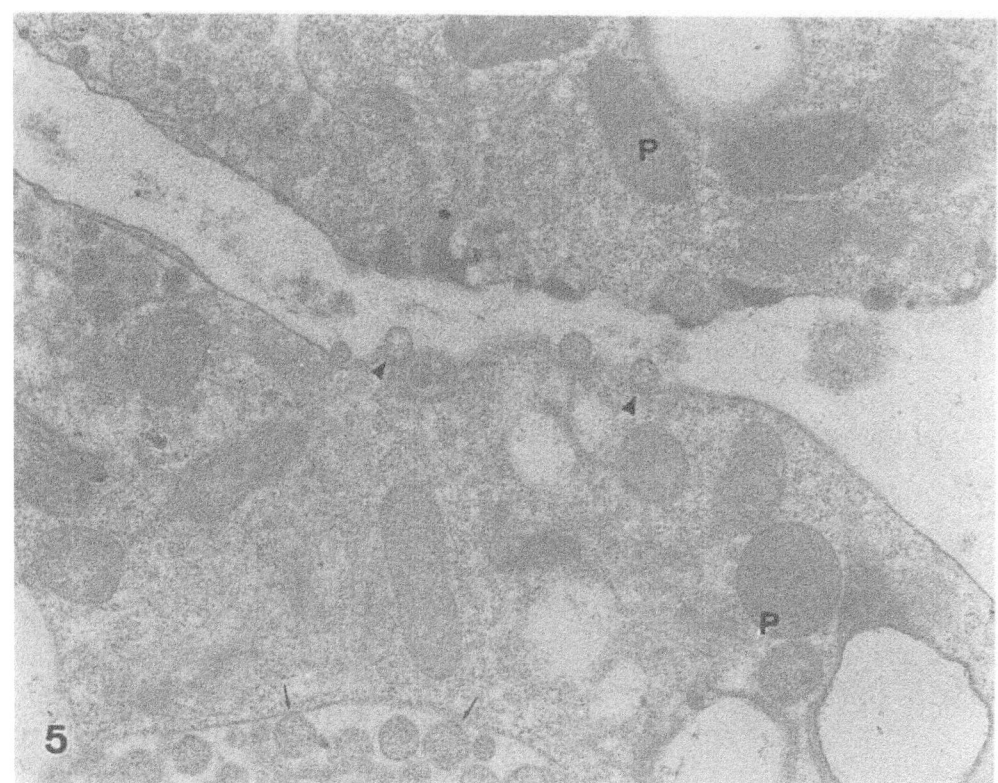

Fig. 5. *Cucurbita pepo.* Parts of two tapetal cells in the middle microspore stage. Small cytoplasmic vesicles containing only ribosomes are visible inside a vacuole (arrows) and are in contact with the plasmamembrane (arrowheads) extruding into the tapetal intercellular space. Note abundance of plastids (*P*). × 15000

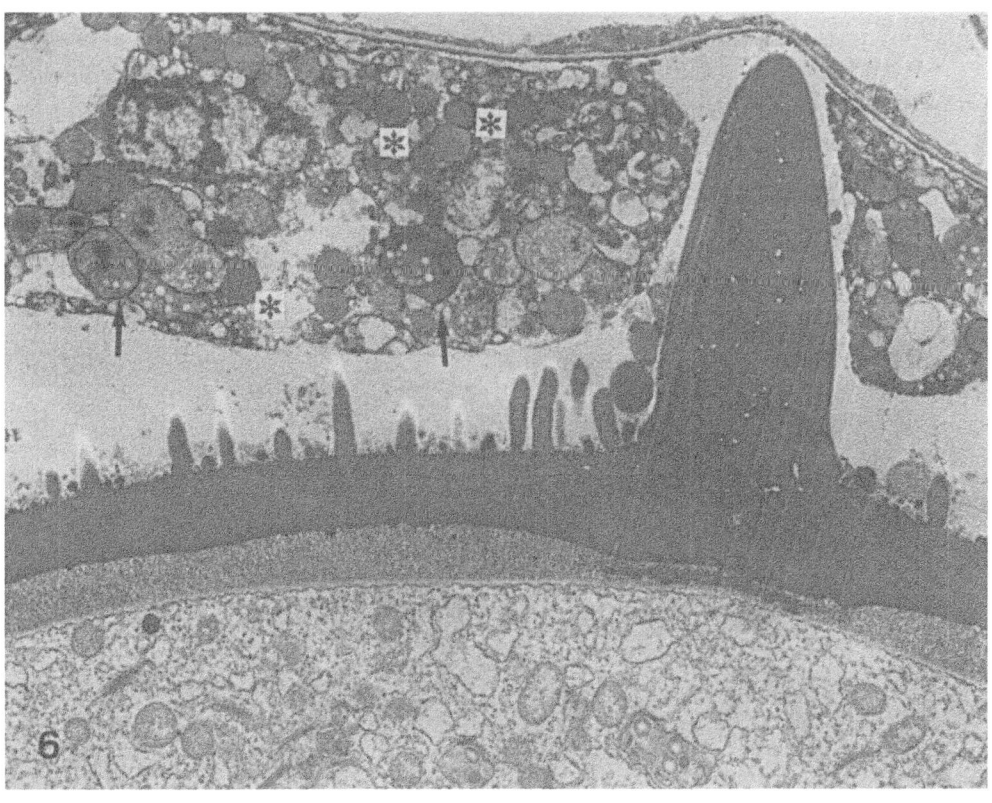

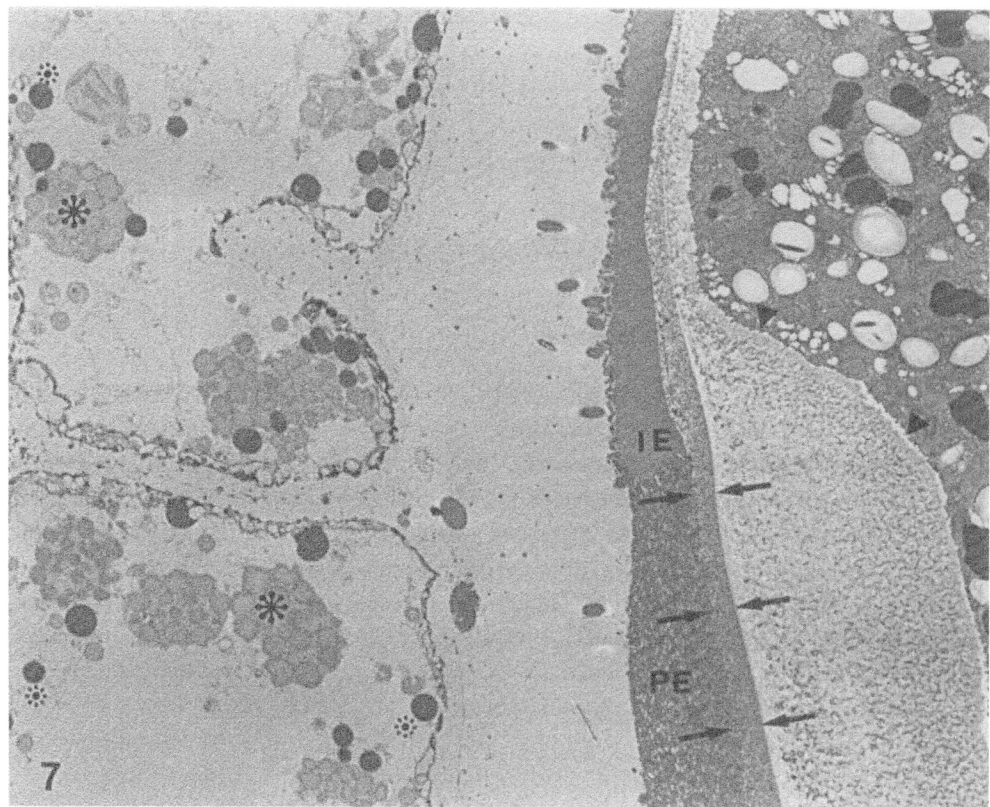

Fig. 7. *Cucurbita pepo*. Parts of two tapetal cells and the pore region of a pollen grain in the early bicellular stage. The interporal exine (*IE*) is compact, the poral exine (*PE*) spongy. A uniform layer of sporophytic proteins is visible between the poral intine and exine (arrows). The poral intine is enlarged and interwoven by tubules where gametophytic proteins are deposited. On the cytoplasmic side of the intine a smooth electron translucent layer is visible (arrowheads). The cytoplasm contains amyloplasts with only one starch grain and highly electron opaque lipid inclusions. In the tapetal cells only elaioplasts with several inclusions (big asterisks) and electron opaque cytoplasmic lipid globules (small asterisks) remain. × 13000

interporal smooth exine. Sporophytic proteins are deposited simultaneously with gametophytic ones localized in the thick poral intine. The enlargement of the loculus occurs more slowly than that of the microspores, and conical exine spines pierce into the tapetal cells (Figs. 1*i* and 6). Tapetum and microspore proplastids differentiate. Microspore plastids show starch granules; tapetal plastids show electron opaque and electron translucent inclusions (Fig. 6). Lipid bodies are observed in the tapetal cytoplasm (Fig. 6).

Fig. 6. *Cucurbita pepo*. Parts of a tapetal cell and a microspore in the late microspore stage. A conical spine reaches into the tapetal cytoplasm. Plastids (arrows) with electron opaque and electron translucent lipid inclusions are in the cytoplasm. Asterisks indicate amyloplasts. × 9500

Early bicellular stage. Tapetal and pollen plastids are already differentiated. Degenerating plastids (formerly elaioplasts) full of lipid inclusions are observed in the tapetal cytoplasm together with strongly electron opaque lipid droplets (Figs. 1*j*, 7).

Middle bicellular stage. The content of the tapetal cytoplasm is reduced to a lipid mass becoming pollenkitt (Fig. 1*k*). The plasmamembrane is no longer visible; only a few electron opaque deposits remain.

Late bicellular stage. Very sticky pollenkitt surrounds the pollen grain (Fig. 1*l*).

Discussion

Pollen grain volume grows continuously from microspore release until pollen dehydration (Pacini 1990b). Tapetal cell volume, increasing during meiosis and tetrad stage, reaches its maximum in the early or middle microspore stage (Pacini & al. 1985); it subsequently decreases until degeneration. Degeneration may be with or without tapetal remains; the remaining substances mostly consist of pollenkitt and are deposited on the pollen grain surface just before anther opening (Pacini & Franchi 1991). In this respect the tapetum of *Cucurbita pepo* is similar to that of many other species: it produces pollenkitt in the same manner as other entomophilous taxa do. A peculiar feature of *Cucurbita pepo* is the reduced size of the tapetal cells with respect to the pollen grains. Normally the pollen diameter is less than the maximum tapetal cell diameter. The contrary is valid for *Cucurbita pepo* which is probably due to the huge size of its pollen grains. Not more than two or three pollen grains are observed in cross sections of the loculus and the need of nutrition may also be correlated with the huge size of the grains: the larger the pollen grain size, the more efficient nutrition must be. When many pollen grains are present in the loculus, the microspores probably move, facilitating the diffusion of substances secreted by the tapetal cells (Pacini & Franchi 1991). In *Cucurbita pepo* this movement is not necessary because so few pollen grains do not compete for nourishment.

Irrespectively of the type of tapetum the tapetal cells are in fact secretory cells. Hence they show one or more of the following features: additional DNA content, two nuclei, polyploidy (D'Amato 1984). In *Cucurbita pepo* they have only one nucleus and this feature too, can be linked to the reduced size of these cells with respect to those of other species. One nucleus may be enough for such an amount of cytoplasm.

Three walls separate the tapetal cytoplasm from the microspore cytoplasm in the early tetrad stage: the tapetal cell wall, the microspore mother cell wall and the callosic wall. The microspore mother cell wall disappears first, followed by the tapetal cell wall. In the late tetrad stage, the tapetal cells have completely lost their walls and the primexine is finished. The tapetal cells can now start their main activities. Substances are formed and are transported by the cells of secretory tapetum and reach the loculus in two different ways: by exocytosis or by transport across the tapetal plasmamembrane. We have only observed exocytosis in a late stage of development when small vesicles of cytoplasm were transported to the loculus.

After disappearing of the callose wall microspore volume grows continuously until the dehydration of the anther and pollen takes place (Pacini 1990b). In

Cucurbita pepo the inner tangential and the proximal part of the radial tapetal cell walls swell and seem to accumulate polysaccharides. These substances, together with those derived from the hydrolysis of the microspore mother cell walls and the callosic walls, are reabsorbed by the developing microspores. Polysaccharides can be stored in tapetal vacuoles, tapetal cell walls, and the locular fluid before being utilized by newly released microspores (DICKINSON 1970, GORI 1982, PACINI & FRANCHI 1983). In *Smilax aspera* (PACINI & FRANCHI 1983) the hydrolytic products of the tapetal cell wall polysaccharides and of the vacuole remain for a while in the loculus before being absorbed by the microspores (PACINI & FRANCHI 1983).

Cell differentiation includes development of plastids. In cells that stores starch or lipids the (pro-)plastids divide and differentiate into amyloplasts or elaioplasts, respectively. Tapetal cell plastids in *Cucurbita pepo* divide actively before they differentiate into elaioplasts. This differentiation is leading to pollenkitt formation similar to that in *Olea europaea* (PACINI & CASADORO 1981) and *Prunus avium* (PACINI & al. 1986). Quite recently it was found that the plastid differentiation is very similar in the anemophilous species *Lolium perenne*: The elaioplasts are reabsorbed by the developing pollen grains (PACINI & al. 1992).

The tapetal cells produce sporophytic proteins which can be deposited inside the exine ornamentations (HESLOP-HARRISON & al. 1973, HESLOP-HARRISON 1975) or on the pore opposite of the gametophytic proteins (PACINI & JUNIPER 1979, PACINI & al. 1981). PACINI & JUNIPER (1979) proposed that poral sporophytic proteins are localized in the pore region. *Cucurbita pepo* pollen grains have poral sporophytic proteins visible under the operculum in the late microspore stage. The exine of the operculum has a spongy structure and can be crossed by the proteins or their precursors from the tapetum. In species like *Urtica dubia*, which has a compact operculum, sporophytic proteins are deposited only in the space between the periphery of the operculum and the annulus (PACINI & al. 1981).

The lifespan of tapetal cells corresponds to the period of pollen development but in taxa such as *Lycopersicum peruvianum* the tapetum, its products, and also its residuals are completely absorbed before the late microspore stage (PACINI & JUNIPER 1984). In other taxa, it disappears later (PACINI & al. 1985). The reason for the long lifespace of the *Cucurbita pepo* tapetum may be the large size of the pollen grains which causes a long developmental time.

References

D'AMATO, F., 1984: Role of polyploidy in reproductive organs and tissues. – In JOHRI, B. M., (Ed.): Embryology of angiosperms, pp. 519–566. – Berlin: Springer.

DICKINSON, H. G., 1970: The fine structure of peritapetal membrane investing the micro-sporangium of *Pinus banksiana*. – New Phytol. **69**: 1065–1068.

GORI, P., 1982: Accumulation of polysaccharides in the anther cavity of *Allium sativum*, clone Piemonte. – J. Ultrastruct. Res. **81**: 1258–1262.

HESLOP-HARRISON, J., 1975: The physiology of the pollen grain surface. – Proc. Roy. Soc. London, B, **190**: 275–299.

– HESLOP-HARRISON, Y., KNOX, R. B., HOWLETT, B., 1973: Pollen wall proteins: "gametophytic" and "sporophytic" functions in the pollen walls of the *Malvaceae*. – Ann. Bot. **37**: 403–412.

HUGHES, J., McCULLY, M. E., 1975: The use of an optical brightener in the study of plant structure. – Stain Technol. **50**: 319–329.

KENRICK, J., KNOX, R. B., 1979: Pollen development and cytochemistry in some Australian species of *Acacia*. – Austral. J. Bot. **27**: 413–427.

O'BRIEN, T. P., McCULLY. M. E., 1981: The study of plant structure – principles and selected methods. – Melbourne: Thermarcarphi Pty.

PACINI, E., 1991a: Tapetum and microspore functions. – In BLACKMORE, S., KNOX, R. B., (Eds): Microspores – evolution and ontogeny, pp. 213–237. – London: Academic Press.

–   1990b: Harmomegathy characters of *Pteridophyta* spores and *Spermatophyta* pollen. – In HESSE, M., EHRENDORFER, F., (Eds): Morphology, development and systematic relevance of pollen and spores. – Pl. Syst. Evol. [Suppl.] **5**: 53–69.

–   CASADORO, G., 1981: Tapetum plastids of *Olea europaea*. – Protoplasma **106**: 289–296.

–   BELLANI, L. M., LOZZI, R., 1986: Pollen, tapetum and anther development in two cultivars of sweet cherry (*Prunus avium*). – Phytomorphology **36**: 197–210.

–   FRANCHI, G. G., 1983: Pollen grain development in *Smilax aspera* L. and possible functions of the loculus. – In MULCAHY, D. L., OTTAVIANO, E., (Eds): Pollen: biology and implications for plant breeding, pp. 183–190. – New York: Elsevier.

–   – 1991: Diversification and evolution of tapetum. – In BLACKMORE, S., BARNES, S. H., (Eds): Pollen and spores – patterns of diversification, pp. 301–316. – Oxford: Syst. Assoc., Clarendon Press.

–   – 1993: Role of the tapetum in pollen and spore dispersal. – Pl. Syst. Evol. [Suppl.] **7**: 1–11.

–   – HESSE, M., 1985: The tapetum: its form, function and possible phylogeny in *Embryophyta*. –Pl. Syst. Evol. **149**: 155–185.

–   JUNIPER, B. J., 1979: The ultrastructure of pollen grain development in the olive (*Olea europaea*). I. Proteins in the pore. – New Phytol. **83**: 157–163.

–   – 1984: The ultrastructure of pollen grain development in *Lycopersicum peruvianum*. – Caryologia **37**: 21–50.

–   FRANCHI, G. G., SARFATTI, G. G., 1981: On the widespread occurrence of poral sporphytic proteins in pollen of dicotyledons. – Ann. Bot. **47**: 405–408.

–   TAYLOR, P. E., SINGH, M. B., KNOX, R. B., 1992: Development of plastids, including amyloplasts and starch granules in pollen and tapetum of rye-grass, *Lolium perenne*. – Ann. Bot. **70**: 179–188.

PEARSE, A. G. E., 1968: Histochemistry-theoretical and applied. I. – London: Churchill.

Addresses of the authors: F. CIAMPOLINI, M. NEPI, E. PACINI, Department of Environmental Biology, Botanical Section, University of Siena, via F. A. Mattioli, I-53100 Siena, Italy.

Pl. Syst. Evol. [Suppl.] 7: 23–37 (1993)

Cycles of hyperactivity in tapetal cells

JOHN R. ROWLEY

Key words: *Pinaceae, Pinus sylvestris, Alismataceae, Echinodorus cordifolius, Nymphaeaceae, Nymphaea colorata, Fagaceae, Querus robur.* – Anthers, tapetum, pollen, secretion.

Abstract: The tapetum is an exceptional plant tissue. As part of differentiation of tapetal cells their walls and plasmodesmata lyse, and then the cells become free to move, to extents specific for taxa. As differentiation continues organelles of tapetal cells show structural changes characteristic of secretory cells, and cell activities are not synchronized. The dedifferentiation and redifferentiation that takes place in tapetal cells is probably rather unusual for plant cells. The highly modified cells become again undifferentiated, plasmodesmata form, and the entire tapetum becomes synchronized. The tapetal cells undergo mitosis without a subsequent cytoplasmic division. A function of the mitoses could be to clean up the genetic code. Plasmodesmata of tapetal cells can be formed without a centrifugally expanding cell plate. It seems that coordination between cells does not occur without plasmodesmata. Tapetal cells together with plasmodesmata form a symplast, as in most plant cells having synchronization of activities.

Those tapetal systems I have studied show a number of cycles of differentiation followed by insertion of plasmodesmata, mitosis, and redifferentiation: *Pinus sylvestris* L. (*Pinaceae*), *Echinodorus cordifolius* (L.) GRISEB. (*Alismataceae*), *Nymphaea colorata* PETER (*Nymphaeaceae*), and *Quercus robur* L. (*Fagaceae*).

PACINI and collaborators present original studies as well as reviewing work involving tapetal cells (e.g., PACINI & JUNIPER 1979, 1983; PACINI & al. 1985; PACINI & KEIJZER 1989; PACINI 1990; PACINI & FRANCHI 1991; CIAMPOLINI & al. 1993). I emphasize works reporting examples involving tapetal cell secretory morphology. They represent conditions similar in appearance to poculiform rER as described by UNZELMAN & HEALEY (1974) for secretory trichomes of *Pharbitis*. The poculiform rER syndrom received this name because during the secretory period a large cup-like storage-vesicle links hypertrophied dictyosomes with a network of dilated rER. Dilations of the poculiform rER system open directly to the cell surface.

Morphologically this is very much like the tapetal cell periods of active secretion we have observed for *Pinus* (WALLES & ROWLEY 1982; ROWLEY & WALLES 1985 a, b, 1987, 1988, 1993). There are other published examples of the cytoplasm of tapetal cells in a poculiform condition which we associate with hyperactive intervals of secretion (e.g., DICKINSON & BELL 1972, 1976 a, b; CARRARO & LOMBARDO 1976 a, b; LOMBARDO & CARRARO 1976; AUDRAN 1979; EL-GHAZALY & NILSSON 1991).

My aim is to show necessity for repeated sampling and, more important, perhaps, development of experimental methods for the analysis of living tapetal cells over a prolonged period of several hours to a day or more.

Material and methods

The plant sample collectioning and the preparation follows the protocols described in ROWLEY & WALLES (1985 a) and ROWLEY & al. (1992 b). Details of fixation protocols are given in the figure texts.

Results

Tapetal cell differentiation in microsporangia of *Pinus sylvestris*. We (WALLES & ROWLEY 1982, ROWLEY & WALLES 1985 a, b, 1987, 1988, 1993) have fixed and examined microsporangia from five trees of *Pinus sylvestris* for nine consecutive years and find that the condition of tapetal cells in relation to cytological stage is very similar each year in spite of climatic conditions that greatly affected rates of stage advance. The stages so far published are documented by nuclear or chromosome characteristics from premeiosis through telophase II and tetrad cell division.

We found that tapetal cells went through intervals of intense secretory activities as judged, for example, by rER dilation, dictyosome vesicles, and high volume densities of vesicles and lipoidal globules. After each active period the cells redifferentiated and became connected via plasmodesmata, undergoing nuclear mitosis but not cytokinesis. Redifferentiation of tapetal cells from a hypersecretory morphology to that similar to young undifferentiated cells was asynchronous until plasmodesmata were formed. Redifferentiation toward hypersecretory form included lysis of plasmodesmata.

The earliest period of hypersecretory morphology in our collections preceded the beginning of meiosis in microspore mother cells (WALLES & ROWLEY 1982: figs. 1–8). By the early stages of meiosis (prior to pachytene) the appearance of tapetal cells resembled that of meristematic cells (WALLES & ROWLEY 1982: figs. 9–16). It should be noted that cells of microsporangia, which include pretapetal cells, show metabolic changes during winter and very early spring (KUPILA-AHVENNIEMI & al. 1978).

---▶

Figs. 1, 2. Sections from early diplotene of meiosis in *Pinus sylvestris*. The endomembrane systems, mainly ER, are greatly dilated and the cytoplasm is unusual in appearance. In Fig. 2 a vesicle lined (star) by a Thiéry-carbohydrate positive membrane opens directly onto the loculus. The cytosol of tapetal cells is packed with ribosomes (see "R" in Figs. 6 and 11). *T* Tapetal cells, *M* microspores, *S* starch in plastids, arrowheads mark mitochondria, arrows dicytosomes, *N* nucleus, *E* endothecial cell. A tapetal marker with associated lipoidal globule is marked by an asterisk. Both figures are portions of micrographs in ROWLEY & WALLES (1988). Fixation: KARNOVSKY's (1965) mixture of glutaraldehyde and paraformaldehyde in 0.06 M phosphate buffer (ph 7.2). Stain: Periodic acid, thiocarbohydrazide, and silver proteinate contrast carbohydrates (THIÉRY 1967). Bars: 1 μm

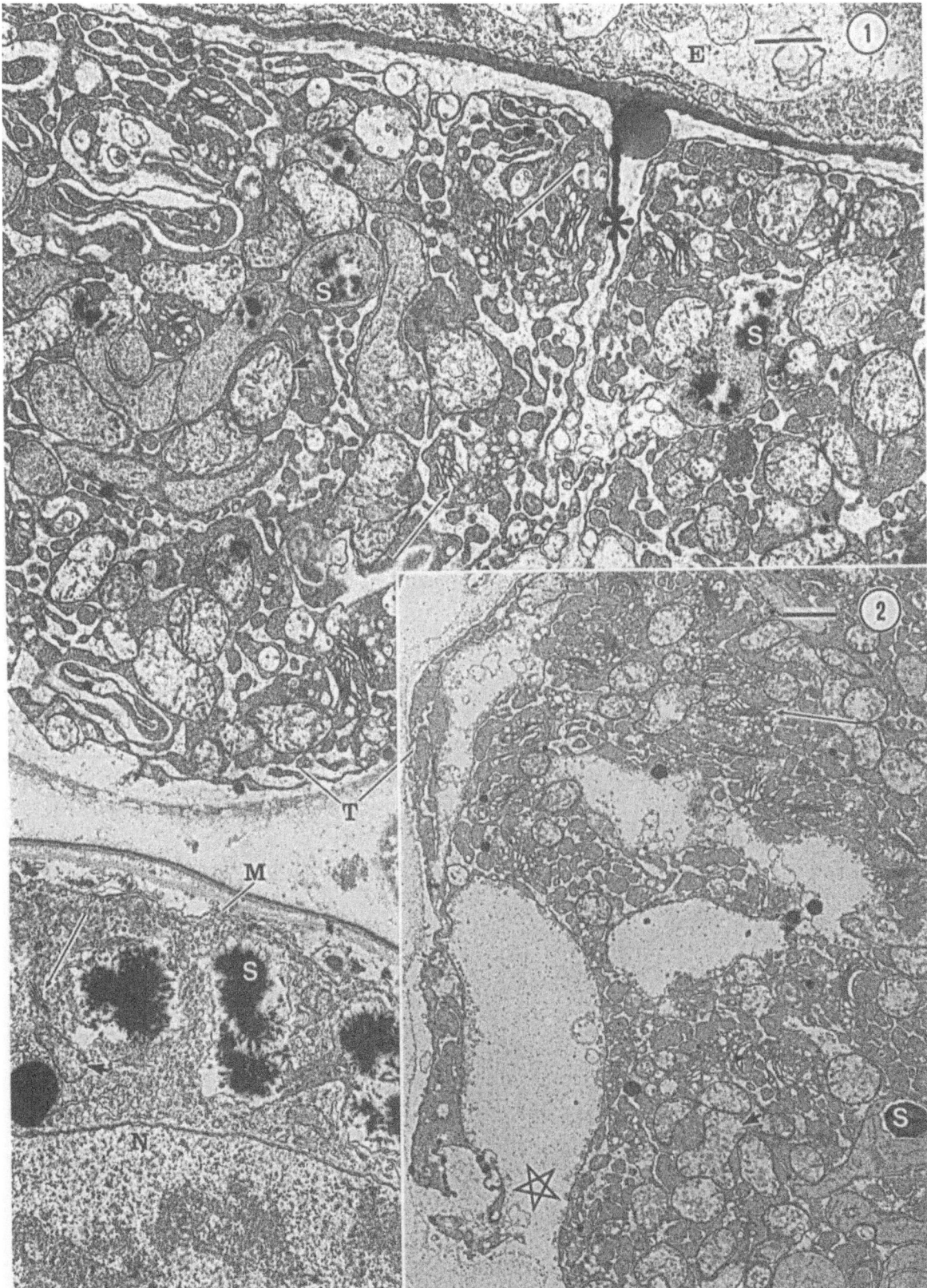

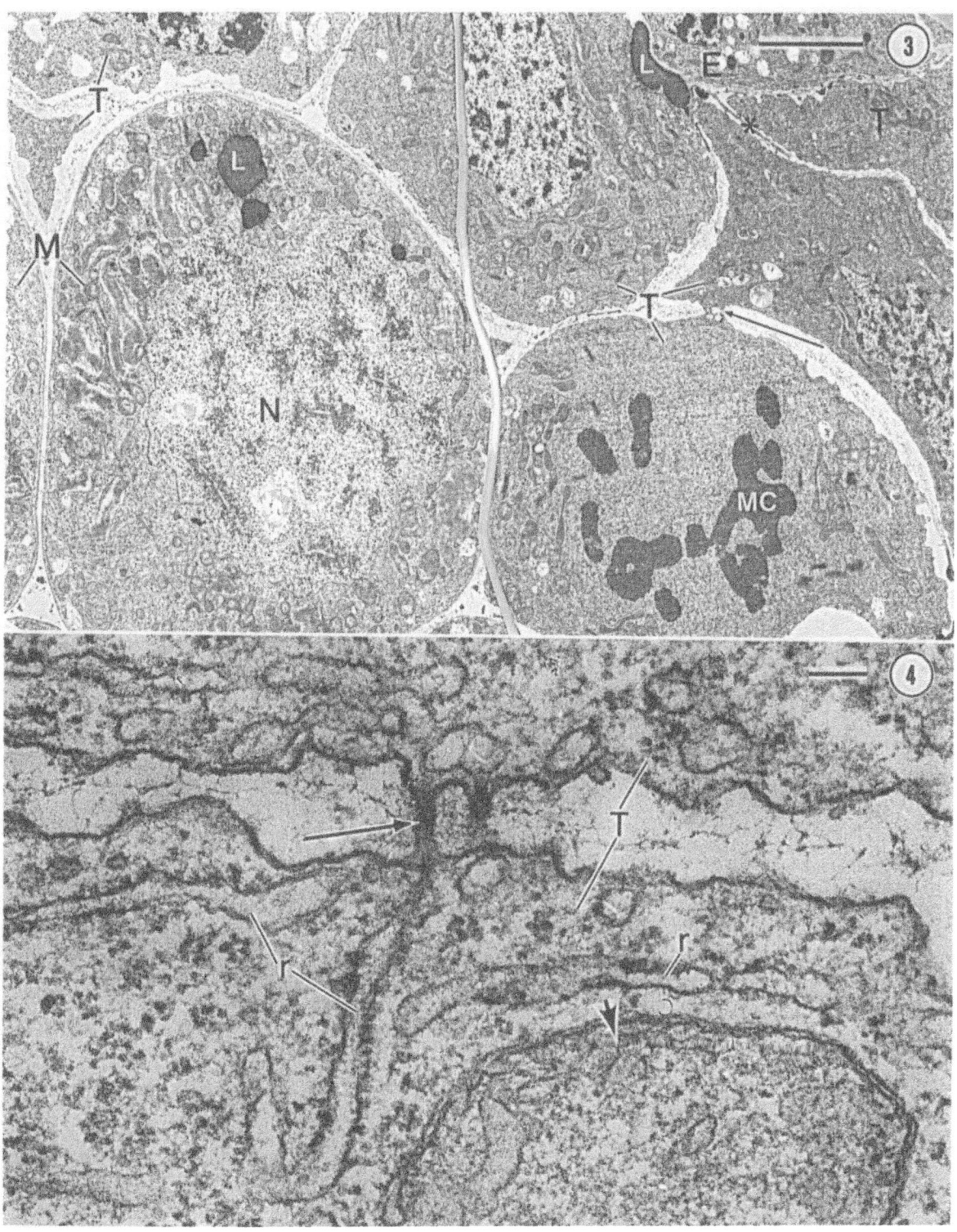

Figs. 3, 4. Early part of the diffuse stage of meiosis in *Pinus sylvestris*. *M* Microspore mother cells; the chromosomes in the nucleus (*N*) are mostly uncoiled. There is no callose enveloping of microspores at this stage. Tapetal cells (*T*) undergo mitosis (*MC* mitotic chromosomes) during this stage and are connected by plasmodesmata (arrow). – Fig. 4. Tapetal cells (*T*). The cisternae of the rER (*r*) are in association with plasmodesmata (arrow). Arrowhead marks mitochondrium. *L* = lipid, *E* = endothecial cell, * = tapetal marker, Fixation: same as Figs. 1 and 2 but with 2% tannic acid and diluted 50% with buffer. Stain: UA-Pb. Bars: Fig. 3: 1 μm, Fig. 4: 0.1 μm

We saw evidence for three distinct periods of hypersecretory morphology during pachytene. Each period was followed by asynchronous recovery, formation of plasmodesmata, and then mitosis. In periods of hypersecretory morphology tapetal cells extended for as much as 80 μm into loculi (ROWLEY & WALLES 1985 a, b).

In the figures presented here I illustrate aspects of two of the cycles of tapetal cell secretory activity and recovery. The first is of an interval of hyperactivity during early diplotene (Fig. 1). Figure 2 shows an aspect of tapetal cells unusual for plant cells—opening of extensive cytoplasmic channels or vesicles directly to the external environment during hypersecretory intervals. The environment of tapetal cells is exceptional since there are no normal cell walls formed within the locule until the pollen grain intine is formed after the mitosis in microspores. The cytoplasm seen at these times of hyperactivity in tapetal cells is unusual but not necrotic (e.g., Figs. 1 and 6.) The cytoplasmic condition of tapetal cells in Fig. 1 had been preceded by mitoses and formation of plasmodesmata (ROWLEY & WALLES 1985 b: figs. 18–20).

The next step within the tapetal cell cycles in *Pinus* is the diffuse stage of meiosis (Fig. 3). The diffuse stage is an interval of chromosome unwinding (cf. KLASTERSKA & RAMEL 1979). Tapetal cells undergo mitosis at this time. Cells of the tapetum are connected by plasmodesmata (Figs. 3 and 4) and during this connection the entire tapetum is composed of young morphologically undifferentiated cells (Fig. 3).

The micrographs in Figs. 5 and 6 illustrate tapetal cell hyperactivity during telophase of meiosis II. The cytoplasm at this time is unusual but healthy in appearance (Fig. 6). Figure 7 shows ER cisternae opening directly into vesicles. The vesicles in Fig. 7 are pro-Ubisch bodies which appear to provide a specialized form of transfer of substances from tapetal cells to the locule and microspores (Ubisch bodies in *Pinus* are illustrated in Fig. 8 and in *Quercus* in Fig. 18).

My final illustrations of tapetal cells in *Pinus* are of an interval during the vacuolate stage in microspore development (Figs. 8 and 9). These figures show that cells of the tapetum were in an undifferentiated condition at this relatively late period of microsporogenesis.

Tapetal cell differentiation in microsporangia of *Quercus robur*. I have observed a series of hypersecretory morphologies in tapetal cells of *Q. robur* (e.g., Figs. 11–13) interspersed between intervals when the cytoplasm of tapetal cells was connected by plasmodesmata and appeared to be like in young undifferentiated cells (e.g., Figs. 10, 15 and 16). The micrographs in Figs. 10 (insert), 14 and 17 show the progression in proexine development during the period covered by Figs. 10–17. Much later in development, during the microspore vacuolate period, tapetal cells are in an undifferentiated condition (Fig. 18). These periods of hyperactivity occurred at similar stages of exine development over three spring seasons.

Hyperactivity in tapetal cells of *Echinodorus cordifolius*. In an on-going study of microspore development in *E. cordifolius*, GAMAL EL-GHAZALY and I find that tapetal cells enter and recover from several periods of activity. The micrograph in Fig. 19 illustrates contiguous tapetal cells with different levels of activity.

Tapetal cell modification in *Nymphaea colorata*. We observed two distinct intervals between the end of meiosis and the vacuolate period in microspore development ot *N. colorata* during which tapetal cells protruded into anther locules (ROWLEY & al. 1992 b). The drawings in Figs. 21, 23 and 24 illustrate these intervals

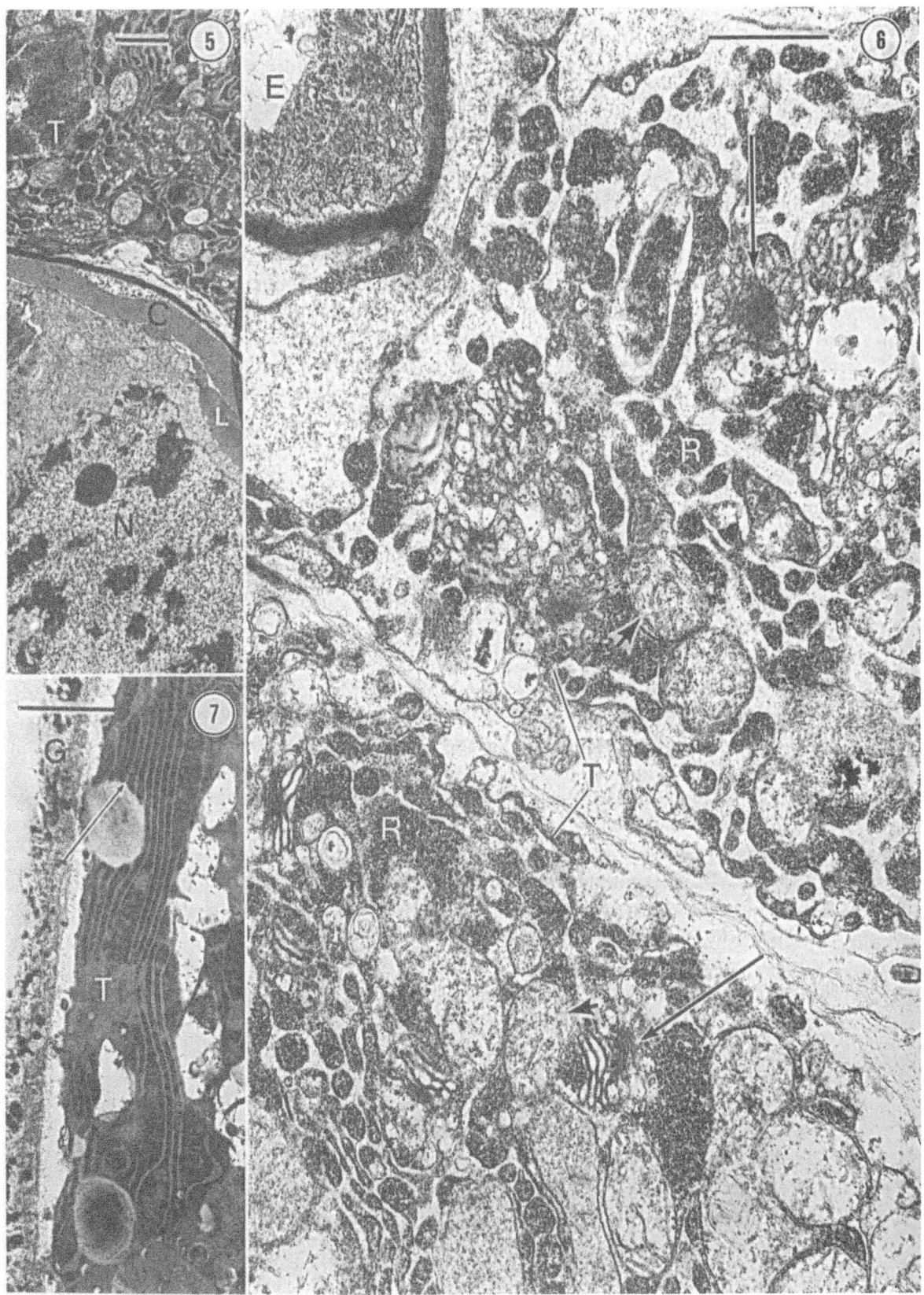

Figs. 5–7. (see legend p. 31)

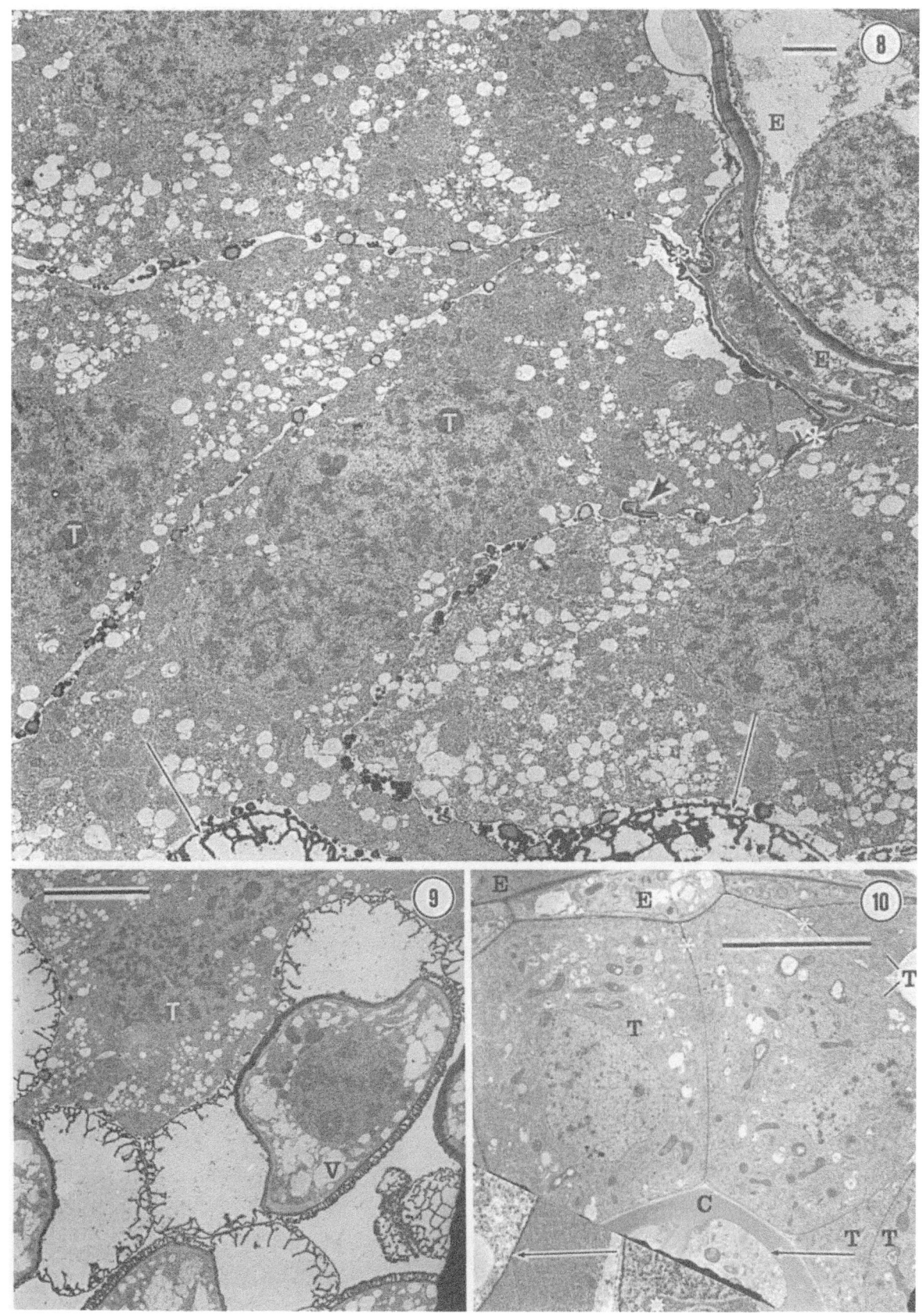

Figs. 8–10. (see legend p. 31)

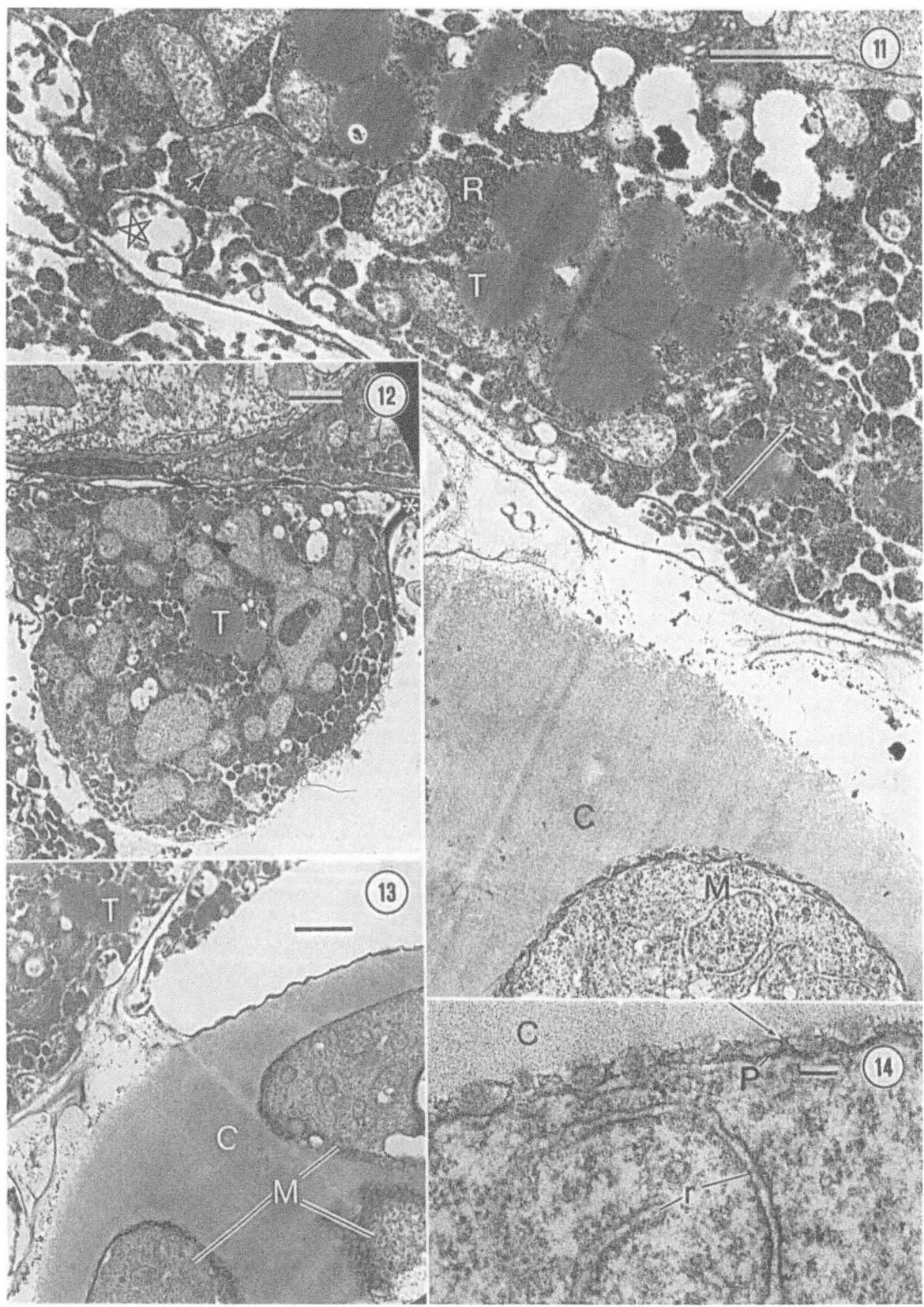

Figs. 11–14

Figs. 5–7. *Pinus sylvestris.* – Figs. 5, 6. Sections from the telophase II period in a micro-sporangium; tapetal cells (*T*) were hypersecretory during this interval. – Fig. 5. *N* Telophase nucleus in the microspore mother cell (within a callosic envelope, *C*). *L* a lipoidal globule within the callosic envelope. – Fig. 6. Portions of the tapetal cells and an endothecial cell (*E*) and its wall. This section was selected because it shows dictyosomes (arrows) in surface and oblique views. The cytoplasm is highly ordered although unusual in appearance because of the enormous area taken by the dilated rER. The cytosol between rER cisternae is dense with ribosomes (*R*). Mitochondria (arrowheads) are prominent in these cells. – Fig. 7. Part of a tapetal cell during metaphase of meiosis. Cisternae of rER (arrow) are contiguous with large vesicles. These vesicles are pro-Ubisch bodies (see ROWLEY & WALLES 1993: Figs. 18–21). *G* Glycocalyx of the tapetal cell. Fixation: Fig. 5. Same as Figs. 1, 2. Stain: UA-Pb; Fig. 6. Same as Figs. 1, 2. Fig. 7. Fixation: Luft's ruthenium red method (LATTA & al. 1975). Stain: UA-Pb. Bars: 1 μm

Figs. 8–10. Microspores. – Figs. 8–9. Early vacuolate stage in microspores of *Pinus sylvestris.* – Fig. 8 includes four tapetal cells (two are marked *T* on nucleoli) each bordered by numerous Ubisch bodies (arrowhead). Ubisch bodies in *Pinus* are formed in several waves and the one marked, with a long "beak", is typical of early productions. Two tapetal markers are distinguished by asterisks; the dark lamellation of these markers and between them is the peritapetal lamellation, separating the loculus from endothecial cells (*E*). Sacci of a microspore are marked by arrows. – Fig. 9. Several microspores with their sacci nestled against a tapetal cell (*T*). – Fig. 10. Early microspore tetrad period in *Quercus robur*. The arrows in the figure and its insert mark the region of the plasma membrane and its surface glycocalyx, the future exine, within the callosic (*C*) envelope. The tapetal cells (*T*) appear undifferentiated. An ameoboid plastid in the insert is marked by a star. *E* Layers of endothecial cells. *V* Vacuole. Figs. 8, 9. Fixation: same as Figs. 1, 2. Stain: UA-Pb. Fig. 10. Fixation: 3% glutaraldehyde plus 2.5% sucrose in 0.1 M phosphate buffer (pH 7.3, 20 C, 24 hr). Stain: UA-Pb. Bars: 1 μm

◀──

Figs. 11–14. A mid-microspore tetrad interval in *Quercus robur*. A period in early exine formation when tapetal cells (*T*) appear to be hyperactive. The proexine is more advanced (e.g., in Fig. 14) than in the stage shown in Fig. 10 and is less advanced than in Fig. 17. – Figs. 11–13 show hyperactive tapetal cells, parts of two endothecial cells (at the top of Fig. 12), and microspores (*M*) within the callosic envelope (*C*). The tapetal cells have greatly dilated rER, dictyosomes (arrows) and a cytosol that is packed with ribosomes (R). A mitochondrium is marked by an arrowhead. The "T"s are located on lipoidal globules. Surface embayments (star) open to the loculus. – Fig. 14. Proexine (arrow) under callose (*C*), microspore plasma membrane (*P*), and rER (*r*). Fixation: same as Fig. 10. Stain: UA-Pb. Bars: Figs. 11–13: 1 μm, Fig. 14: 0.1 μm

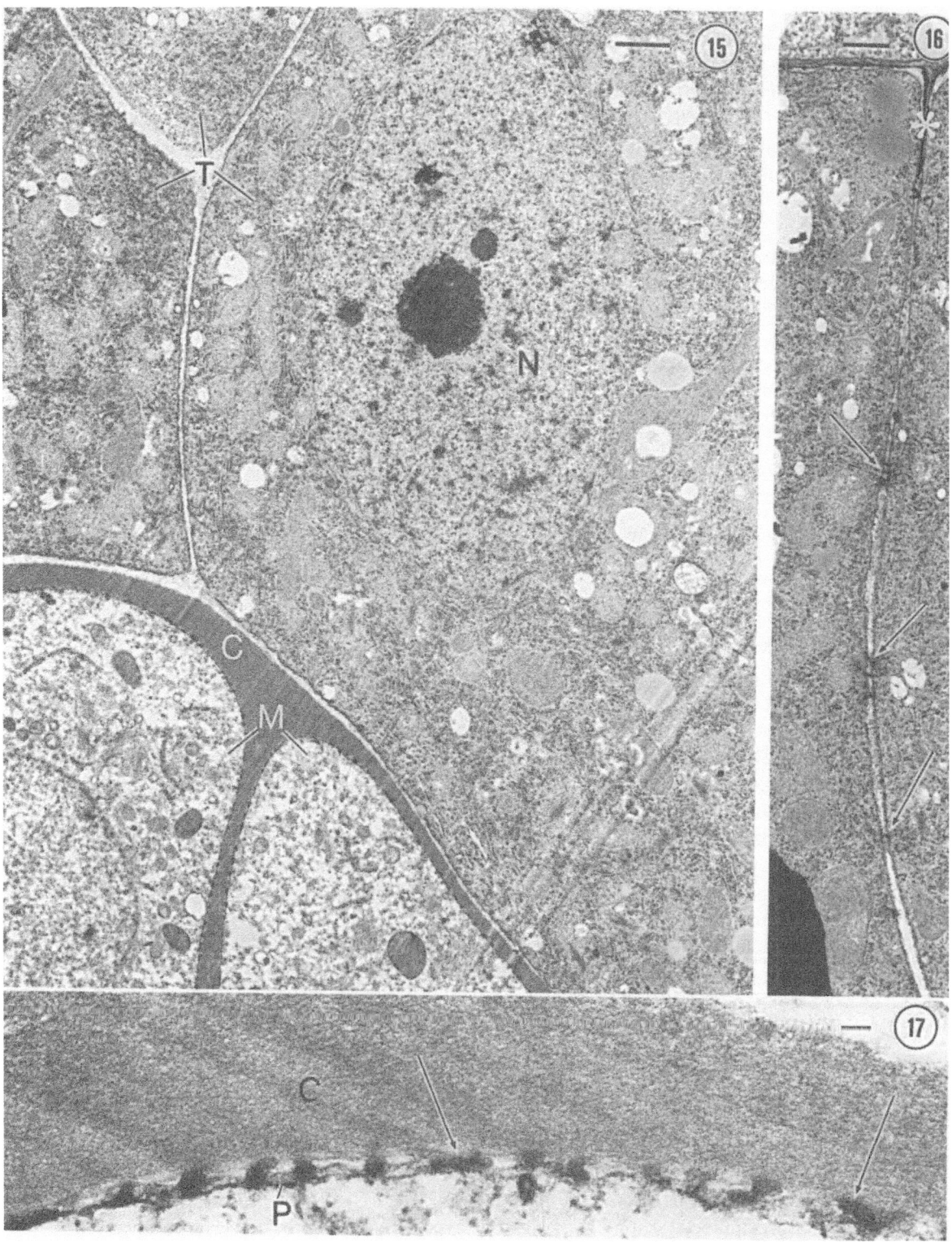

Figs. 15–17. Late microspore tetrad period of *Quercus robur*. – Fig. 15. Portions of two microspores (*M*) and the callosic wall (*C*) below tapetal cells (*T*), *N* tapetal cell nucleus. – Fig. 16. Plasmodesmata (arrows) between tapetal cells from the same section as Fig. 15. There is an endothecial cell at the top of Fig. 16 and a tapetal marker (asterisk) between tapetal cells. – Fig. 17. A micrograph of the proexine on the microspores in Fig. 15: *P* plasma membrane, *C* callosic envelope. The proexine glycocalyx units (arrows) appear to be coiled. Fixation: same as Fig. 10. Stain: 0.1% phosphotungstic acid in 10% chromic acid (considered to contrast polysaccharides). Bars: Figs. 15–16: 1 μm, Fig. 17: 0.1 μm

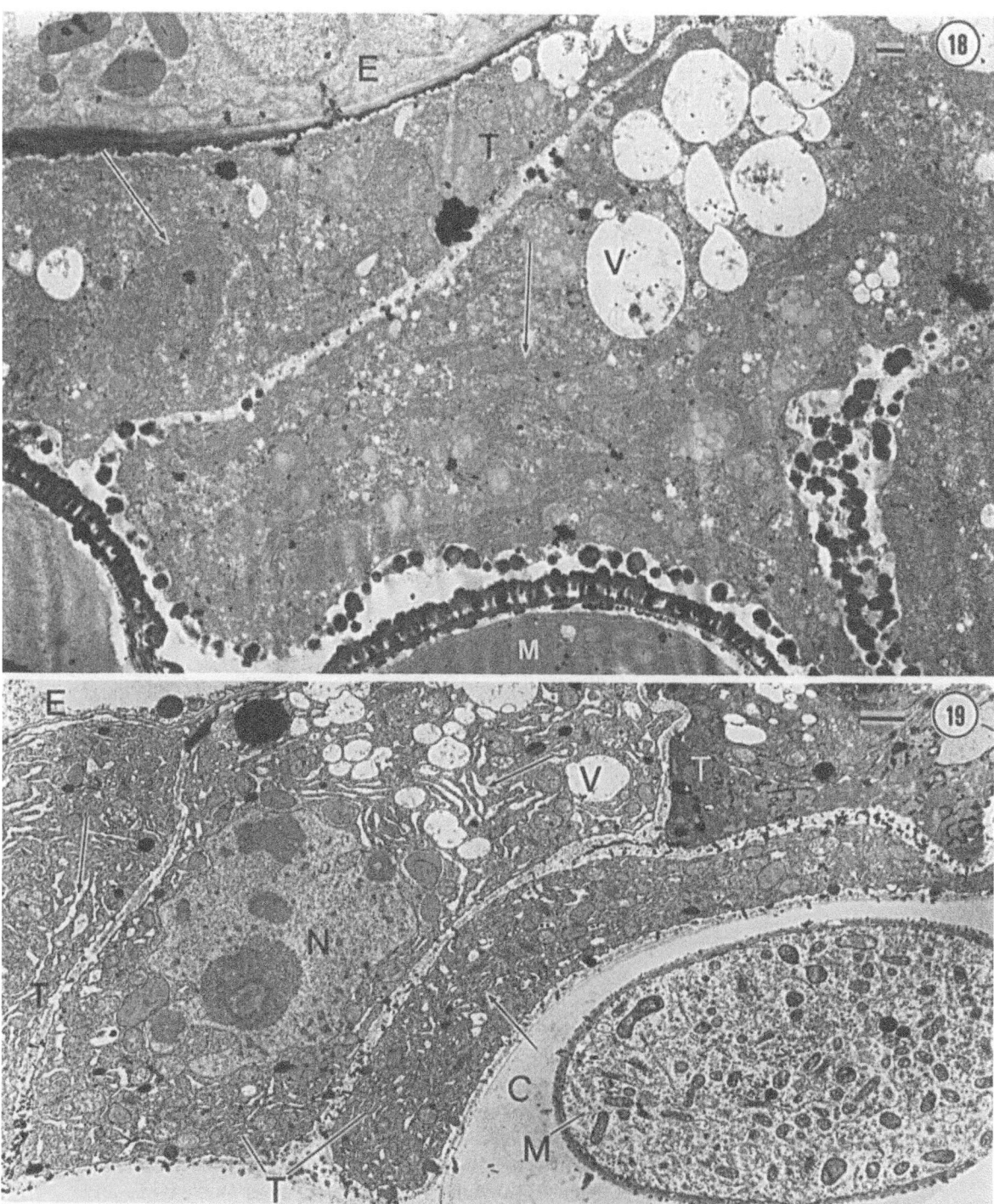

Figs. 18, 19. Microspores. – Fig. 18. Vacuolate stage in *Quercus robur*. The microspores (*M*) have a well developed exine. Tapetal cells (*T*) contain many vesicles (*V*) but other cytoplasmic features, e.g., undilated ER (arrows) do not indicate a high level of activity. Tapetal cells are bordered by Ubisch bodies. *E* Endothecial cell at the top. – Fig. 19. Microspores of *Echinodorus cordifolius* in a late tetrad stage. During this interval tapetal cells asynchronously take on what appears to be a hyperactive secretory condition. Here the four tapetal cells (*T*) differ in the extent of ER (arrows) dilation and in the frequency of vesicles (*V*) and lipoidal globules (dark in contrast). The microspore tetrad (*M*, one microspore in figure) is enveloped by callose (*C*). The nuclei (*N*) of tapetal cells have a fimbriate surface (an indication of an active state). *E* Endothecial cell. Fig. 18. Fixation: same as Fig. 10. Stain: 5% phosphotungstic acid in 10% acetone (considered to contrast proteins). Fig. 19. Fixation: 1% glutaraldehyde plus 1% lanthanum nitrate, then to osmium tetroxide plus 1% lanthanum. Stain: UA-Pb. Bars: 1 μm

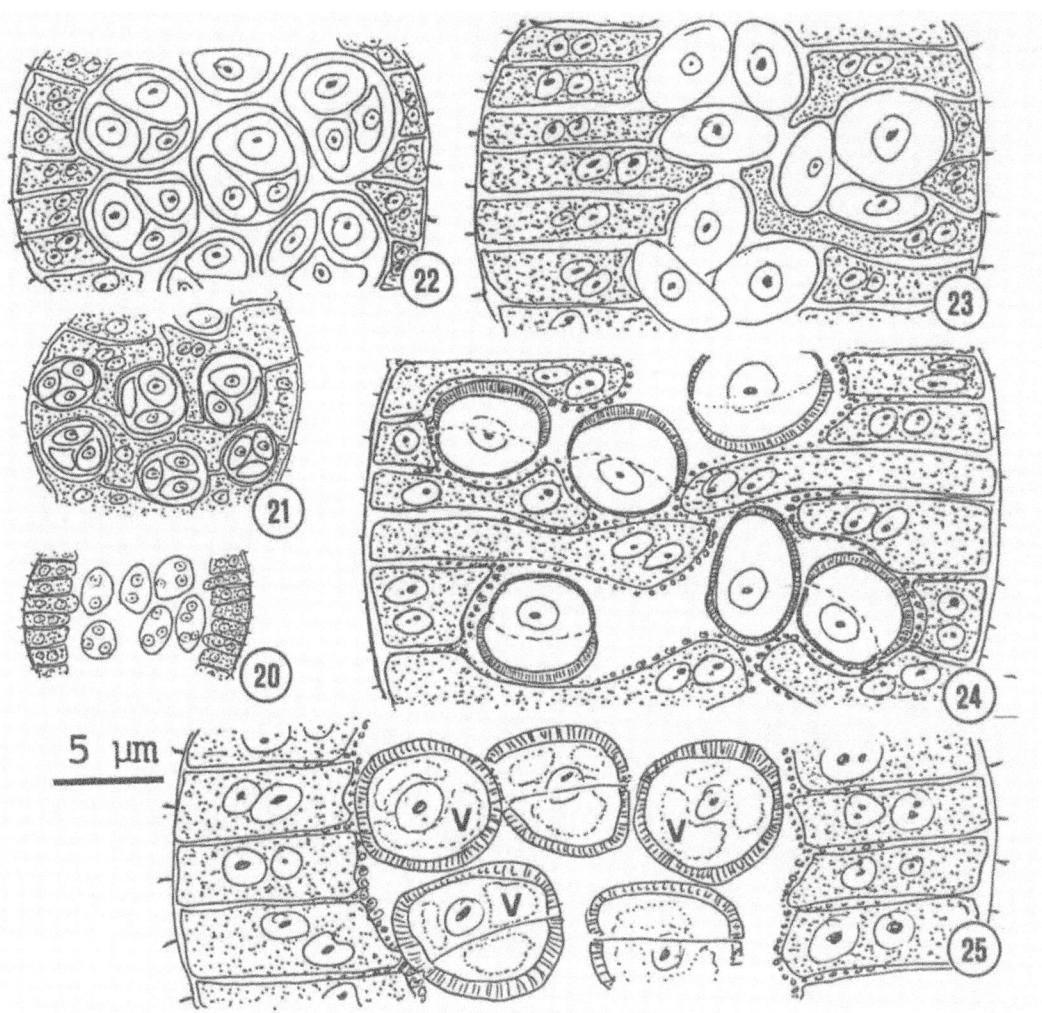

Figs. 20–25. Cyclic modifications in tapetal cells of *Nymphaea colorata*. The sketches are traced from photomicrographs of anther locules. Tapetal cells are stippled. – Fig. 20. Telophase II of meiosis; tapetal cells are arranged in a uniform palisade. – Fig. 20. Early microspore tetrads. Tapetal cells extend into locules and each tetrad is surrounded by tapetal cells; some microspores are adjacent to parietal cells. – Fig. 22. Early proexine. Microspore tetrads are nested in tapetum but tapetal cells are not invasive. – Fig. 23. Microspores are becoming separated. Tapetal cells extend into loculus. – Fig. 24. Microspores are free and have a relatively thick exine. Tapetal cells extend as "bridges" across locules. – Fig. 25. Vacuolate interval in microspore development; *V* vacuole. Tapetal cells arranged in a noninvasive palisade. Ubisch bodies coat locular facing surface of tapetal cells in Figs. 24 and 25

of tapetal cell invasion into locules. Before and after tapetal cell invasions the tapetum resembles an ordered palisade of cells (Figs. 20, 22, 25); in ROWLEY & al. (1992 b) we show TEM-micrographs of tapetal cells and corresponding microspores for each stage in these drawings.

Discussion

Do tapetal cells increase in number?

The invasion of tapetal cells in *Nymphaea* fills the space between microspores during the tetrad period. Later, with greatly increased locular size, invading tapetal cells appear like bridges (Fig. 24). This great increase in size of tapetal cells and locules is evident in Figs. 20–25 and documented by hundreds of embryological studies, many of which were reviewed by DAVIS (1966), MAHESHWARI (1950), and PERIASAMY & SWAMY (1966).

The question, however, is, can the original complement of tapetal cells increase sufficiently to provide for microspores in the greatly enlarged locules? While there are many observations of mitoses in tapetal cells and much has been written about endomitosis in tapetal cell nuclei I know of no published reports on cell division in the tapetum. The reason why tapetal cells do not increase in number may be that higher plant cells do not divide without cell wall formation (e.g. ROBINSON 1971). The tapetum, at least in angiosperms and gymnosperms, is surrounded by a sporopolleninous layer termed the e x t r a t a p e t a l   m e m b r a n e by HESLOP-HARRISON (1969), the peritapetal membrane by DICKINSON (1970). HESLOP-HARRISON (1969) considered the lamellation to be the boundary of a "culture sac" since it becomes sporopolleninous and limits portions of the pollen locules or microsporangia containing tapetal cells and microspores. When we began work with *Pinus* (WALLES & ROWLEY 1982) and saw the prominent remnants of tapetal cell walls extending into loculi from endothecial cell walls we referred to these remnants as "tapetal markers" thinking that in mature microsporangia they would indicate the original position of tapetal cells. This supposition may be interesting, but considering the relatively vast increase in the volume of microsporangia and size increases of tapetal cells the idea of a "tapetal marker" should be reexamined. In nine years of study of *Pinus sylvestris*, however, we have seen no evidence indicating autolysis of components of the peritapetal lamellation; rupture of the peritapetal lamellation is presumably a prerequisite for tapetal cell replacement from endothecial cells. We have not seen any indication of division of tapetal cells, and it appears that they are capable of enormous increases in area/volume.

Inadequacy of fixed cells in assessing dynamic events and processes. What the future would have been of cells examined after fixation can not be known by seeing cells killed during later stages in the ontogeny. The cyclic modifications in tapetal cells indicate a dramatic ontogeny which can hardly be adequately understood by periodic sampling. The examples I present exemplified activities of tapetal cells but do not resolve them. An urgent need is continuous observations of living materials like those of OLGA ERDELSKÁ (1983) for the embryo sacs of *Galanthus nivalis* and *Jasione montana* using Nomarski contrast and time-lapse microcine-photography. My substitute has been to observe comparable stages in development year after year in *Pinus* and *Quercus*.

In the study of *Pharbitis* by UNZELMAN & HEALEY (1974) the active stage of secretion, the poculiform condition, was followed by senescence of the secretory cells. In tapetal cells a hyperactive period of secretion may be followed by mitosis, formation of plasmodesmata and a general redifferentiation with cells then

appearing morphologically like young undifferentiated cells. Since tapetal cells –
like the microspores they help support – differ from other plant cells, they are
promising subjects for experimental studies.

In much published work on tapetal cells the first sign of hypersecretory activity
seems to be interpreted as the beginning of tapetal cells degeneration and senescence.
While I appreciate the limited life expectancy of tapetal cells I find it illogical to
propose that support cells of the heterotrophic microspores/pollen grains are
withdrawn before pollen grains are almost fully formed and nutritionally independ-
ent. It looks to me as though tapetal cell systems of many taxa undergo similar, but
asynchronous, cycles of secretary activity followed by redifferentiation.

I thank EDEL ALSTERBORG, TUULIKKI LINDQVIST, and SUSANNE LINDWALL for technical
assistance. My work was supported by grants from the Swedish Research Council. Figures
1–3 are portions of figures published in ROWLEY & WALLES (1988). Figure 5 is serial with
a section in ROWLEY and WALLES (1993). Figure 7 is part of a micrograph published in
ROWLEY & WALLES (1987). Figures 20–25 are rearranged and slightly revised from ROWLEY
& al. (1992 a).

References

AUDRAN, J.-C., 1979: Microspores, pollen grains and tapetum ontogeny in *Ceratozamia
mexicana* (*Cycadaceae*): An ultrastructural study. – Phytomorphol. **29**: 350–362.

CARRARO, L., LOMBARDO, G., 1976 a: Tapetal ultrastructural changes during pollen development. II. Studies on *Pelargonium zonale* and *Kalanchoë obtusa*. – Caryologia **29**: 339–344.

– – 1976 b: Tapetal ultrastructural changes during pollen development. III. Studies on
Gentiana acaulis. – Caryologia **29**: 345–349.

CIAMPOLINI, F., NEPI, M., PACINI, E., 1993: Tapetum development in *Cucurbita pepo*
(*Cucurbitaceae*). – Pl. Syst. Evol. [Suppl.] **7**: 13–22.

DAVIS, G. L., 1966: Systematic embryology of the angiosperms. – New York: Wiley.

DICKINSON, H. G., 1970: The fine structure of a peritapetal membrane investing the
microsporangium of *Pinus banksiana*. – New Phytol. **69**: 1065–1068.

– – BELL, P. R., 1972: The rôle of the tapetum in the formation of sporopollenin-containing
structures during microsporogenesis in *Pinus banksiana*. – Planta **107**: 205–215.

– – 1976 a: Development of the tapetum in *Pinus banksiana* preceding sporogenesis. – Ann.
Bot. **40**: 103–113.

– – 1976 b: The changes in the tapetum of *Pinus banksiana* accompanying formation and
maturation of the pollen. – Ann. Bot. **40**: 1101–1109.

EL-GHAZALY, G. A., NILSSON, S., 1991: Development of tapetum and orbicules of
Catharanthus roseus (*Apocynaceae*). – In BLACKMORE, S., BARNES, S. H., (Eds): Pollen and
Spores. – Syst. Assoc. Spec. Vol. **44**: 317–329. – Oxford: Clarendon Press.

ERDELSKÁ, O., 1983: Microcinematographical investigations of the female gametophyte,
fertilization and early embryo- and endosperm development. – 7th Internat. Symp. on
Fertilization and embryogenesis in ovulated plants. pp. 49–52. High Tatra, Rackova
dolina, Czechoslovakia. – Bratislara: VEDA.

HESLOP-HARRISON, J., 1969: An acetolysis-resistant membrane investing tapetum and
sporogeneous tissue in the anthers of certain *Compositae*. – Canad. J. Bot. **47**: 541–542.

KARNOVSKY, M. J., 1965: A formaldehyde-glutaraldehyde fixative of high osmolarity for use
in electron microscopy.–J. Cell Biol. **27**: 137A.

KLASTERSKA, I., RAMEL, C., 1979: Prophase of plant meiosis: sequences and interpretation
of stages. – Genetica **51**: 15–20.

KUPILA-AHVENNIEMI, S., PIHAKASKI, S., PIHAKASKI, K., 1978: Wintertime changes in the ultrastructure and metabolism of the microsporangiate strobili of the Scotch pine. – Planta **144**: 19–29.

LATTA, H., JOHNSTON, W. H., STANLEY, T. M., 1975: Sialoglycoproteins and filtration barriers in the glomerular capillary wall. – J. Ultrastruct. Res. **51**: 354–376.

LOMBARDO, G., CARRARO, L., 1976: Tapetal ultrastructural changes during pollen development. I. Studies on *Antirrhinum majus*. – Caryologia **29**: 113–125.

MAHESHWARI, P., 1950: An introduction to the embryology of angiosperms. – New York: McGraw-Hill.

PACINI, E., 1990: Tapetum and microspore function. – In BLACKMORE, S., KNOX, R. B., (Eds): Microspores: evolution and ontogeny. – London: Academic Press.

–   JUNIPER, B. E., 1979: The ultrastructure of pollen grain development in the Olive (*Olea europaea*). 2. Secretion by the tapetal cells. – New Phytol. **83**: 165–174.

–   – 1983: The ultrastructure of the formation and development of amoeboid tapetum in *Arum italicum* MILLER. – Protoplasma **117**: 116–129.

–   KEIJZER, C. J., 1989: Ontogeny of intruding non-periplasmodial tapetum in the wild chicory, *Cichorium intybus* (*Compositae*). – Pl. Syst. Evol. **167**: 149–164.

–   FRANCHI, G. G., 1991: Diversification and evolution of the tapetum. – In BLACKMORE, S., BARNES, S. H. (Eds): Pollen and spores. – Syst. Assoc. Spec. Vol. **44**, pp. 301–316. – Oxford: Clarendon Press.

–   FRANCHI, G. G., HESSE, M., 1985: The tapetum: its form, function, and possible phylogeny in *Embryophyta*. – Pl. Syst. Evol. **149**: 155–185.

PERIASAMY, K., SWAMY, B. G. L., 1966: Morphology of the anther tapetum of angiosperms. – Curr. Sci. **35**: 427–430.

ROBINSON, D. G., 1991: What is a plant cell? The last word. – Pl. Cell **3**: 1145–1146.

ROWLEY, J. R., WALLES, B., 1985 a: Cell differentiation in microsporangia of *Pinus sylvestris*. II. Early pachytene. – Nordic J. Bot. **5**: 241–254.

–   – 1985 b: Cell differentiation in microsporangia of *Pinus sylvestris*. III. Late pachytene. – Nordic J. Bot. **5**: 255–271.

–   – 1987: Origin and structure of Ubisch bodies in *Pinus sylvestris*. – Acta Soc. Bot. Poloniae **56**: 215–227.

–   – 1988: Cell differentiation in microsporangia of *Pinus sylvestris*: Diplotene and the diffuse stage. – Ann. Sci. Nat. Bot. (13) **9**: 1–28.

–   – 1993: Cell differentiation in microsporangia of *Pinus sylvestris*: Diakinesis through cytokinesis. – Nordic J. Bot. **13**: 67–82.

–   GABARAYEVA, N. I., WALLES, B., 1992 a: Tapetal cell modifications around microspores of *Nymphaea*. – XI Internat. Symp. on embryology and seed reproduction, pp. 467–468. – St. Petersburg: NAUKA.

–   – 1992 b: Cyclic invasion of tapetal cells into loculi during microspore development in *Nymphaea colorata* (*Nymphaceae*). – Amer. J. Bot. **79**: 801–808.

THIÉRY, J.-P., 1967: Mise en évidence des polysaccharides sur coupes fines en microscopie électronique. – J. Microscopie **6**: 987–1018.

UNZELMAN, J. M., HEALEY, P. L., 1974: Development, structure, and occurrence of secretory trichomes of *Pharbitis*. – Protoplasma **80**: 285–303.

WALLES, B., ROWLEY, J. R., 1982: Cell differentiation in microsporangia of *Pinus sylvestris* with special attention to the tapetum. I. The pre- and early-meiotic periods. – Nordic J. Bot. **1**: 53–70.

Address of the author: JOHN R. ROWLEY, Botany Department, University of Stockholm, S-106 91 Stockholm, Sweden.

Pl. Syst. Evol. [Suppl.] 7: 39–52 (1993)

Pollenkitt development and composition in *Tilia platyphyllos* (*Tiliaceae*) analysed by conventional and energy filtering TEM

MICHAEL HESSE

Key words: *Tilia, Tiliaceae.* – Tapetum, microsporogenesis, pollenkitt, elaioplasts, endoplasmic reticulum, EELS, ESI.

Abstract: Advanced TEM techniques (improved chemical fixation protocols and new instrumental TEM features) can give new insight into old problems. Pollenkitt, the main cause of pollen stickiness in entomophilous angiosperms, was shown especially by thin-layer chromatography to be an inhomogeneous mixture of many, mostly neutral lipids. The production of pollenkitt in the tapetum starts after meiosis in the microspore stage. In *Tilia platyphyllos* besides plastids other organelles contribute substantially to pollenkitt components; this was proved by the employment of adequate fixation techniques, especially osmium ferrocyanide ("Os-FeCN") used as a postfixative. SER-connected lipid droplets represent one type of pollenkitt precursors, and elaioplasts produce the second lipid/carotenoid pollenkitt precursor component. Both components are composed of inhomogeneous droplets with differing electron density. Mature pollenkitt adhering to the pollen is depicted by Conventional TEM (CTEM) mostly as a widely homogeneous, rather compact substance. Differing electron densities indicating variations in the chemical composition within pollenkitt droplets can only sometimes be visualized at CTEM level. An Energy Filtering Transmission Electron Microscope (EFTEM) improves by enhanced grey level discrimination the specific contrasts of lipid pollenkitt droplets. If comparing the sample in the zero loss imaging with images taken by contrast tuning this technique allows the finding of variations in electron density originating from variable chemical composition; so image analysis of lipid/carotenoid droplets is permitted. Thus at the CTEM level it is confirmed that also seemingly homogeneous droplets are structurally indeed inhomogeneous.

Mature pollen grains of zoophilous (and amphiphilous) angiosperms are generally coated with an oily, sticky material, commonly called pollenkitt. Pollenkitt formation was well documented in the past (e.g., REZNICKOVA & DICKINSON 1982, KEIJZER 1987) but this was evidently not the final word with respect to organelle contribution. Pollenkitt is produced in the anther tapetum. Mostly the plastids – and often also the tapetal cytoplasm and other organelles – are involved in its formation (WEBER 1992b). It forms the mostly dense and structurally homogeneous exine coating as seen by light microscopy and also by conventional TEM (ROWLEY & EL-GHAZALY 1992). Investigations of mature and/or degenerating tapetal cells by TEM (REZNICKOVA & DICKINSON 1982), and especially by thin-layer chromatography

(Dobson 1988) have shown that pollenkitt varies considerably interspecifically and is a complex mixture of many unsaturated and saturated lipids, of carotenoids, and sometimes also of proteins.

Pollinating insects are attracted by a specific pollination syndrome including form, colour and odour of the flowers. The odour results from nectaries and/or from the pollen and from the pollen coat. The mostly neutral pollenkitt lipids belong to the essential, volatile oils. Pollenkitt lipids might therefore be the vector of pollen aromas, which would indicate that pollenkitt function is not restricted to make pollen more or less sticky. It indeed provides the pollen grains with species-specific oils and acts as an olfactory attractant especially to pollen-seeking insects. Certain constituents of pollenkitt can be used as recognition markers by flower visiting insects, i.e. as an olfactory attractant to pollinators (Henning & Teuber 1992). But if pollenkitt lipids are indeed highly diverse, the fundamental question arises: is all this diversity merely the result of the necessity to attract insects? Or do some pollenkitt components serve other, so far less known or widely neglected purposes, e.g., as protection against fungi or other parasites?

The structural diversity in pollenkitt composition could so far not be proved also in early developmental stages using classical fixation protocols and Conventional Transmission Electron Microscopy (CTEM). The present paper shows that advanced TEM techniques can clearly demonstrate that pollenkitt substances originate from more than a single tapetum organelle and are in fact a structurally complex mixture, consisting of a cocktail of several substances.

Material and methods

Anthers of *Tilia platyphyllos* Scop. (*Tiliaceae*) were fixed in 3% Sörensen- or Karnovsky-buffered glutaraldehyde for 24 h at room temperature. After rinsing in buffer and distilled water postfixation was either carried out in 2% aqueous OsO_4 for 2 h, or in a 1:1 mixture of 2% aqueous OsO_4 and 0,8% phosphate buffered $K_3Fe(CN)_6$ (osmium ferrocyanide, "*Os-FeCN*"), for 4 h (Weber 1992a, b). After washing in distilled water the anthers were dehydrated in an ethanol series and embedded in Spurr's Low Viscosity Resin. For CTEM, ultrathin sections were either conventionally stained by uranyl acetate and lead citrate or tested for insoluble polysaccharides or for neutral lipids (Weber 1992a) and viewed in a Conventional Transmission Electron Microscope (a ZEISS EM 109 or EM 900).

Sections of *Tilia platyphyllos* and – in comparison – of *Alisma plantago-aquatica* L. were viewed in the Energy Filtering Transmission Electron Microscope (EFTEM, a ZEISS EM 902). Either extremely thin, unstained sections were investigated in the brightfield/zero loss as well as the filter imaging mode, or thicker sections of conventionally stained material were viewed in the filter mode. The latter was done by selecting that part of the electron energy loss spectrum, where the contrast is optimized, i.e. using the contrast tuning technique and the plasmon loss technique (for review of these EELS-ESI techniques and applications see Reimer & al. 1991; Fromm & al. 1992; and Fehrenbach & al. 1991, 1992).

Results

There is evidence for pollenkitt production by more than a single tapetum organelle. In *Tilia platyphyllos* two different types of pollenkitt (pk) precursors, showing separate cytological and/or compartimental origins, occur within the tapetum cells.

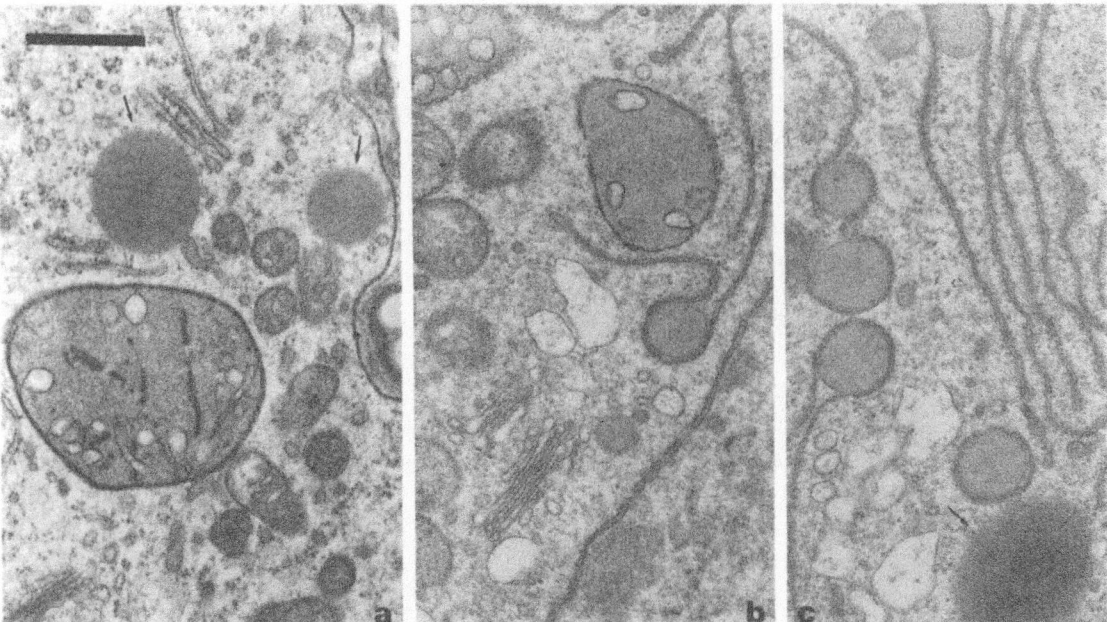

Fig. 1. *Tilia platyphyllos. a* Tapetum during late tetrad stage. Lipid droplets (arrows) within the cytoplasm, representing the first type of pollenkitt (pk) precursors. Proplastids present (undifferentiated). *b, c* Tapetum during free microspore stage after callose dissolution. Lipid droplets (first pollenkitt precursor type) are surrounded and enclosed by ER-profiles, or not (in *c,* arrow). Proplastids present (undifferentiated). – Bar: 0,5 μm

The first – initially not membrane-bound – lipid droplets within the cytoplasm can be found as early as in the late tetrad stage when callose still surrounds the young microspores (Fig. 1 a). After testing for neutral lipids they exihibit a faint, highly electron dense outermost layer, a coating. Later on, in the free microspore stage, during exine formation, the number of these globules increases, and they become tightly enclosed by SER-profiles (Fig. 1b, c). When exine formation is more or less completed, the globules – representing one pollenkitt component – are released into the cytoplasm, and are no longer enclosed by ER. Tapetal plastids do not show lipid droplets at this stage. Prior to the first microspore mitosis the number and dimension of the formerly SER-bound globules (the first pk precursor type) are greatly enlarged; only few connections to ER profiles can be seen at this stage (Figs. 2a, 3). After the first microspore mitosis the abundant non-plastidal, cytoplasmic lipid droplets often appear to be again connected with or even enclosed by a single (SER-)membrane (Figs. 4a, b, 5). The cytoplasmic pk precursor globules have varied electron density, which can already be seen in the brightfield mode, and show either electron-transparent spaces (Fig. 5) or dense inclusions within a highly transparent core (Figs. 2 a, 4b, 6 a). Only by contrast tuning the globules reveal further, otherwise masked differences in electron density (Fig. 6 a–c): The detectable number of dense globules using an EFTEM varies considerably depending on the actual spectrum of the electron energy loss mode.

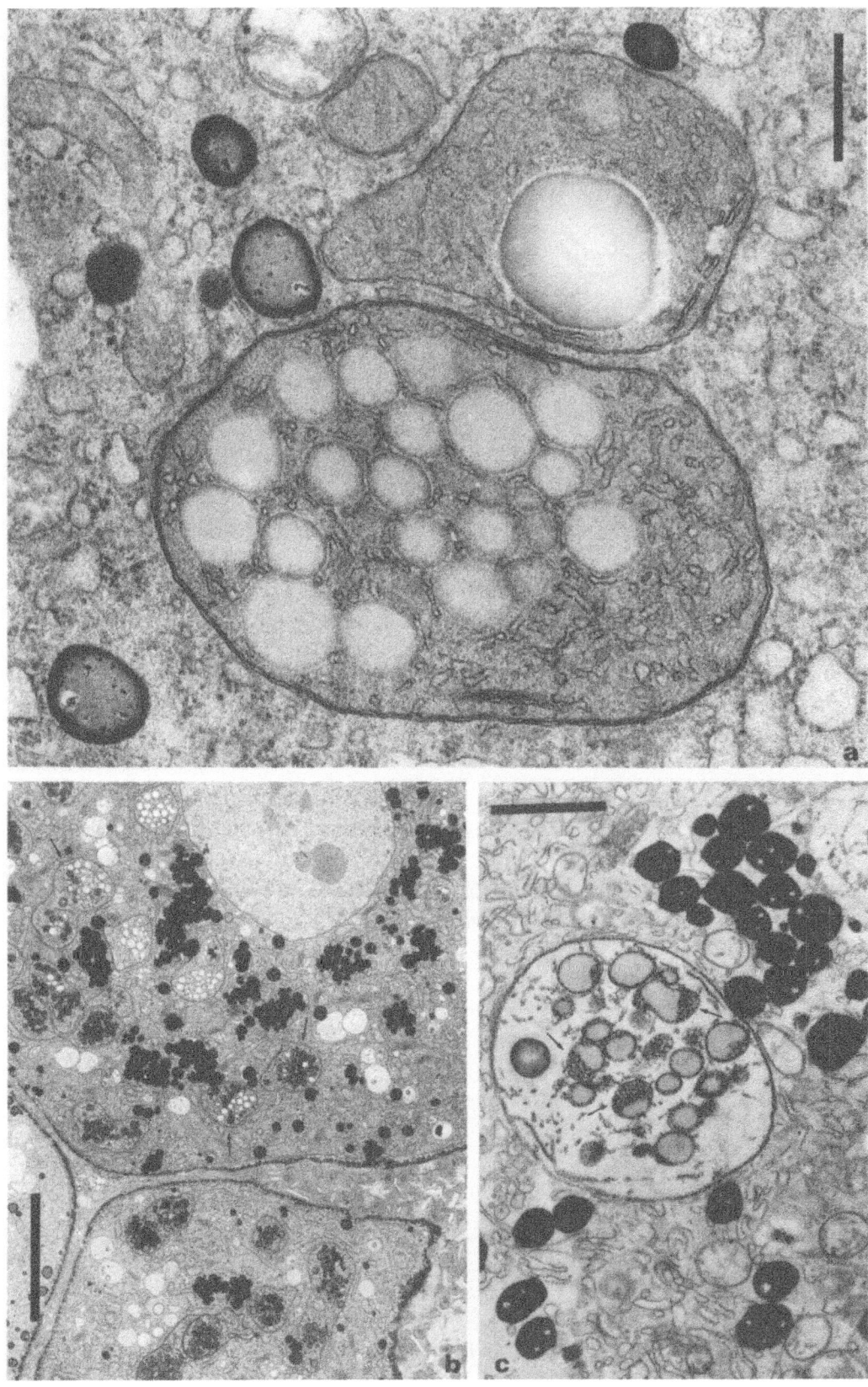

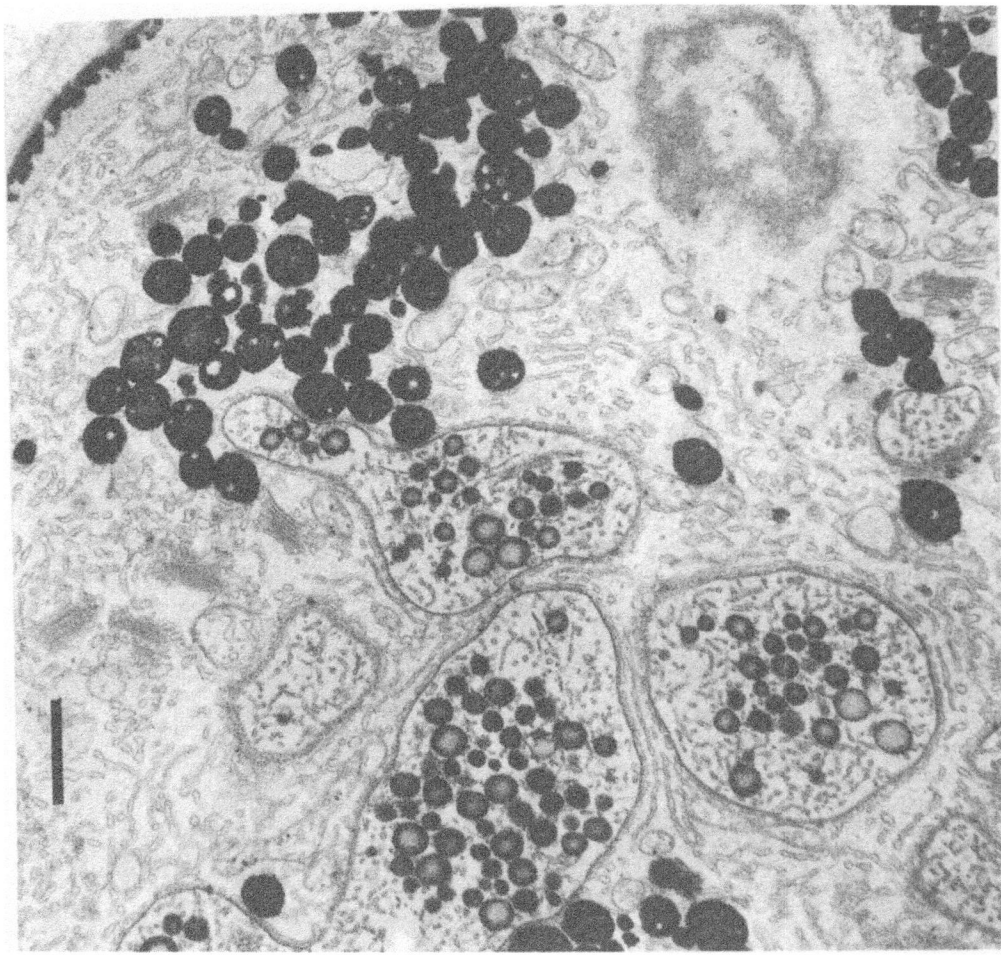

Fig. 3. *Tilia platyphyllos*. Tapetum at first haploid mitosis. The plastid pk droplets consist of a medium dense core and an extremely dense outer layer. The non-plastid, "cytoplasmic" pk component is very rarely connected with ER-profiles in this developmental stage. Section tested for neutral lipids. Bar: 1 μm

◀ —————————————————————————————

Fig. 2. *Tilia platyphyllos*. *a* Tapetum at late microspore stage. Some droplets of the first pk precursor type lie free within the cytoplasm. Modified plastids contain few starch grains and many electron transparent lipid droplets. Bar: 0, 5 μm. *b, c* Shortly after the developmental stage shown in *a:* The first pk type now accumulates as highly electron dense globules in the cytoplasm. The plastids contain many lipid droplets with quite different density (arrows in *b*), many droplets become structurally inhomogeneous (bubbles, arrows in *c*). *b* Section tested for neutral lipids, Bar: 5 μm, *c* section THIÈRY-tested, Bar: 1 μm

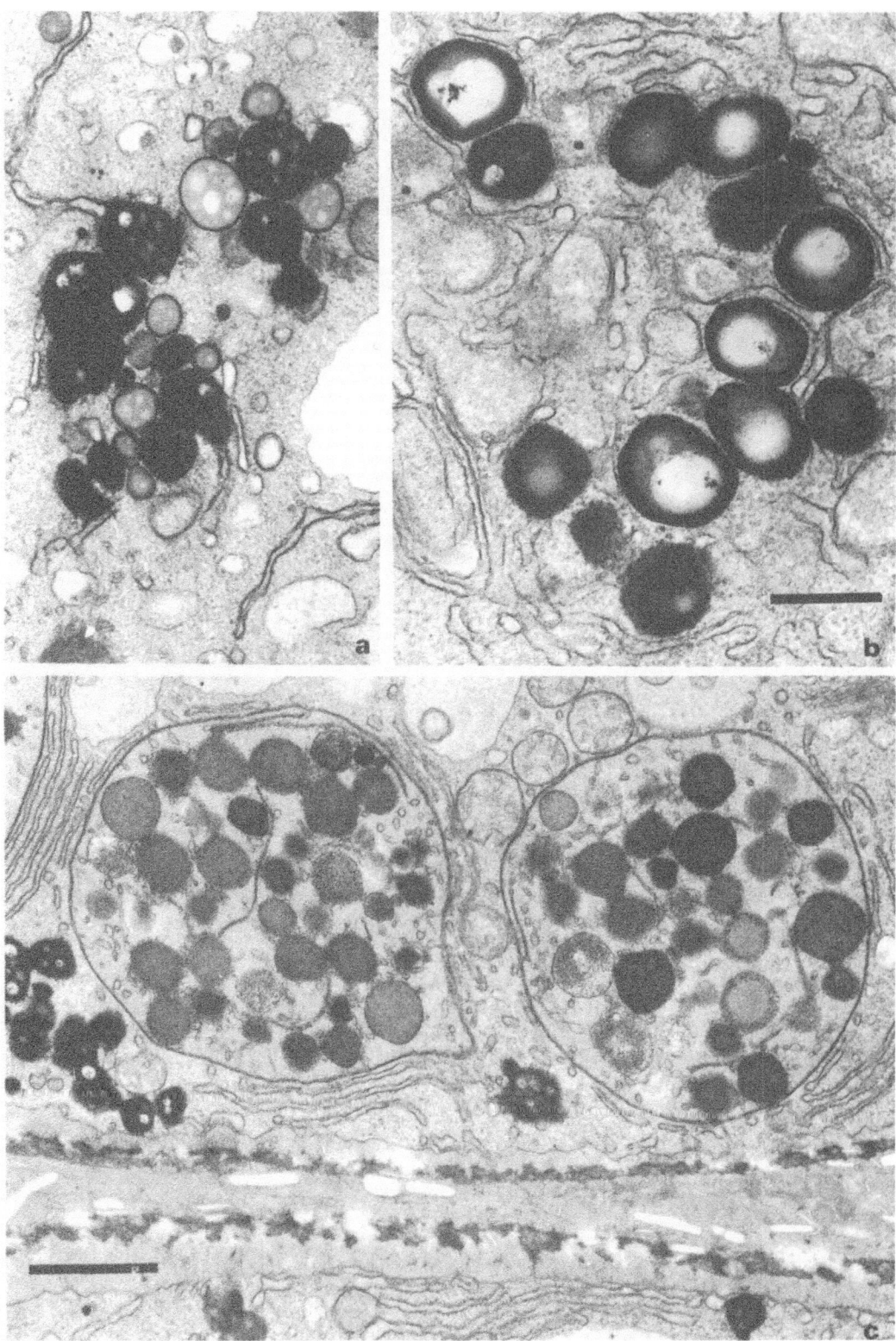

Fig. 4. *Tilia platyphyllos*. Tapetal cells. After first microspore mitosis, pollen grain intine formation is nearly completed. *a, b* At this stage of development the globules of the first pk precursor type appear to be more frequent surrounded by ER-profiles. They differ highly in their electron-density: often spaces or electron-transparent cores are seen. Bar: *a, b:* 0,5 μm. *c* Outer and inner plastid membrane in close proximity leaving a seemingly single plastid membrane. Note the different electron density and structure of the second (plastid) pk precursor type. Bar: 1 μm

The second pk component is formed independently from the first one within modified tapetal plastids (elaioplasts). Up to exine formation stage the plastids do not show modifications and do not produce any lipid droplets. Shortly before the first pollen mitosis the plastids within the tapetal cells undergo remarkable modifications. Only very few thylakoids are formed, but many proplastid-like tubules can be seen. The plastids (elaioplasts) enlarge greatly and form a lot of widely electron-transparent lipid droplets. Shortly after the first pollen mitosis, also the plastidal droplets undergo some significant changes. They show bubbles (Fig. 2c) and a striking difference in electrondensity. Some elaioplasts are filled with highly electron-transparent globules, some with electron-dense globules, while others contain both types simultaneously (Fig. 2 b). Shortly after this stage the plastidal droplets show an electron-transparent core and a dense outer zone (Fig. 3). The pk globules are composed of a granular component also, which derives from the

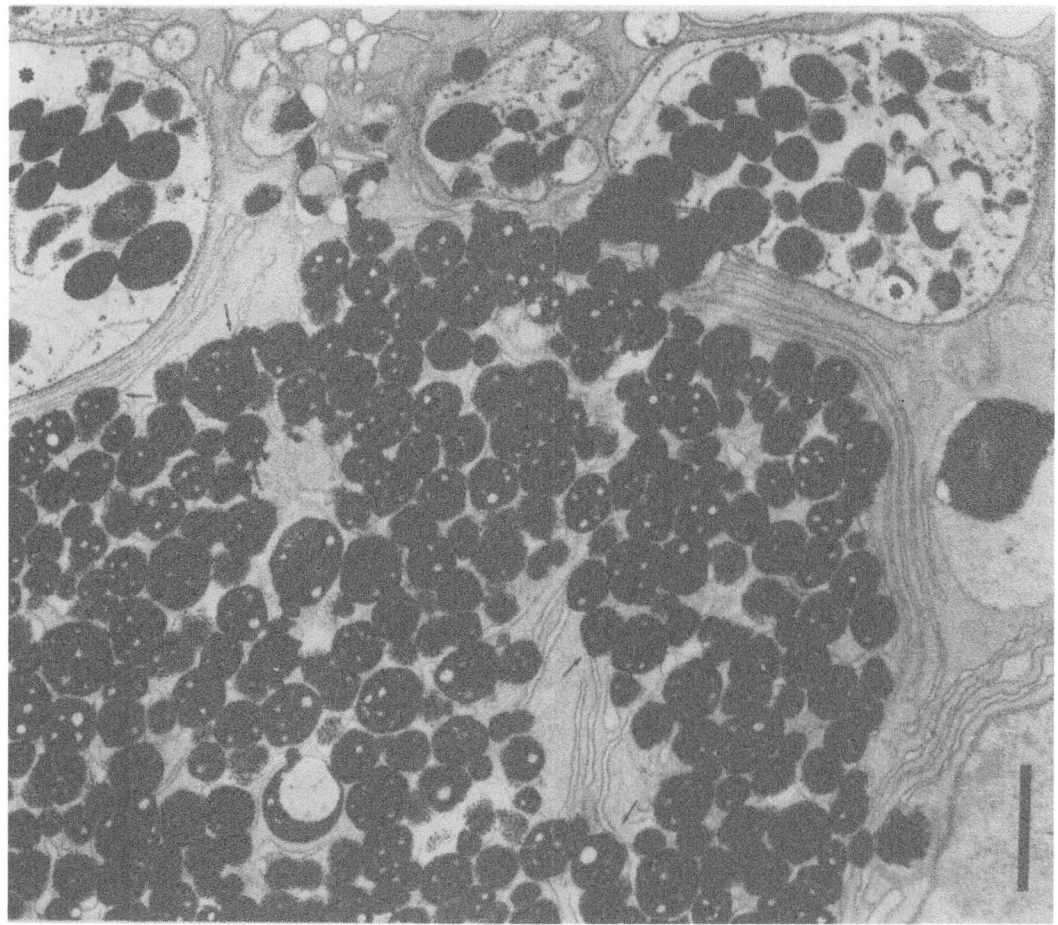

Fig. 5. *Tilia platyphyllos*. Tapetal cell shortly after the developmental stage shown in Fig. 4c, pollen grain intine formation completed. The globules of the non-plastidal pk component appear surrounded or even connected with SER-profiles (arrows), and the droplets get more electron transparent areas. The plastid pk component shows more and more its dual composition of extremely dense and transparent lipid areas (asterisks). Bar: 1 μm

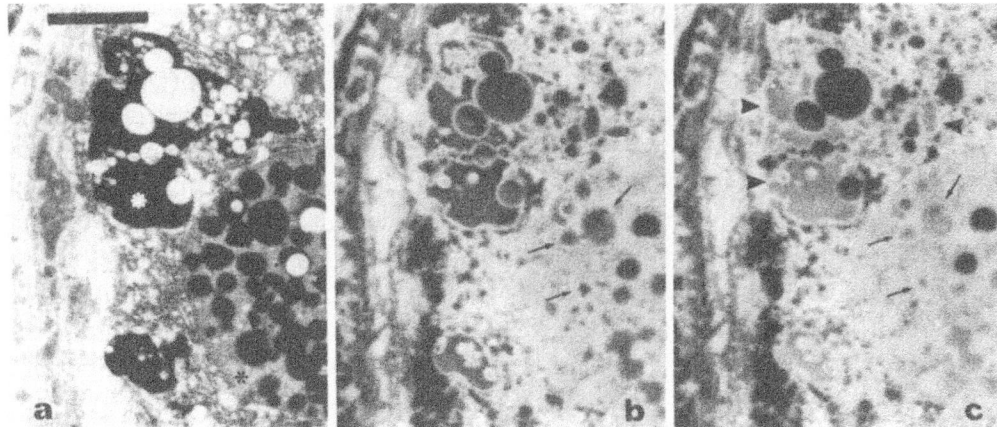

Fig. 6. *Tilia platyphyllos*. Tapetum just before degeneration begins (mature pollen grain stage). Both types of pk precursors seen in an Electronic Filtering TEM in zero loss mode (*a*), and the very same sample in the contrast tuning mode (*b* and *c*), respectively. The elaioplasts (black asterisk) contain – as shown in the zero loss mode – many electron-dense and -transparent droplets; both filter images (*b, c*) show (depending on the contrast tuning technique) a different number of globules (arrows). Similarly, the lipid droplets (arrowheads) of the first, the non-plastidal pk precursor type (white asterisk) also demonstrate different electron densities in the filtered images compared with the zero loss image *a* indicating different compositions. *a* $\Delta E = 0$ eV, *b* $\Delta E = 88$ eV, *c* $\Delta E = 131$ eV. Bar: 1 µm

globular bubbles seen in an earlier stage (Figs. 2 c, 4 c). Furthermore, as in the case of the non-plastidal, SER-connected droplets, the plastidal droplets differ not only in structure and electron-density, which can be seen in the brightfield mode (Fig. 5), but also in a character, which cannot be detected in the brightfield mode of a conventional TEM. Viewed in an electronic filtering TEM in the zero loss mode, the plastid droplets show only very few grey level discrimination, so that the extremely electron-dense lipid material appears to be strictly homogeneous. But viewed in the electron energy loss mode (by contrast tuning), however, seemingly homogeneous droplets exhibit otherwise masked zones of different electron density (Fig. 6 a–c). That this phenomenon is not only restricted to *Tilia platyphyllos* and early developmental stages is demonstrated by Fig. 7, comparing pk droplets of *Alisma plantago-aquatica* in the zero loss and the contrast tuning modes.

Until late stages of tapetum development in *Tilia platyphyllos* both pk precursor types are well separated. Only during the final degeneration stages of the tapetum cells the second pk precursor type becomes released from the plastids, as now the plastid membranes degenerate. Both precursor types independently float into the loculus. First they can be distinguished as blistered lumps occurring from the SER-born component and as small, mostly dense droplets resulting from the plastidal pk precursor type. Later on both components may become mixed. The components are finally deposited on the pollen grain exine. They are partly situated on top of the exine tectum or are filling the intercolumellar spaces either as small droplets or – at the latest stage just before dehiscence – as a widely compact, sometimes lamellated, but generally extremely dense substance, forming the proper pollenkitt (Fig. 8).

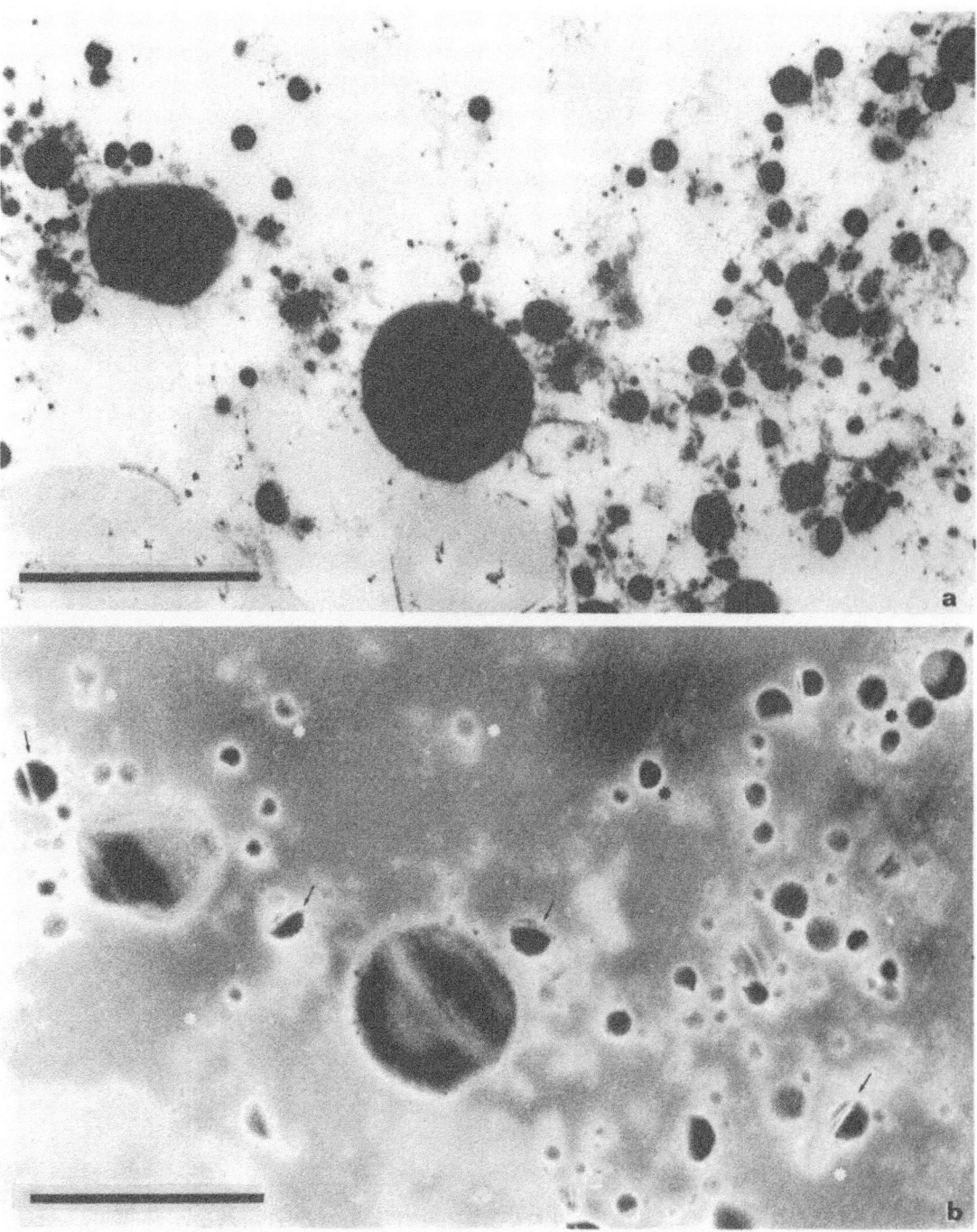

Fig. 7. *Alisma plantago-aquatica.* After tapetum degeneration is completed pk droplets float between mature pollen grains, seen by EFTEM in zero loss (*a*), and in the contrast tuning mode (*b*). *a* Lipid pk droplets with various dimensions, but of generally high electron-density. The filtered image (*b*) shows within (in *a*) seemingly homogeneous large or small droplets (arrows) different electron densities. Moreover, some electron-dense globules appear as transparent in the filtered image (white asterisks), while others retain their dense habit more or less unchanged (black asterisks). *a* $\Delta E = 0\,eV$, *b* $\Delta E = 97\,eV$. Bar: 1 μm

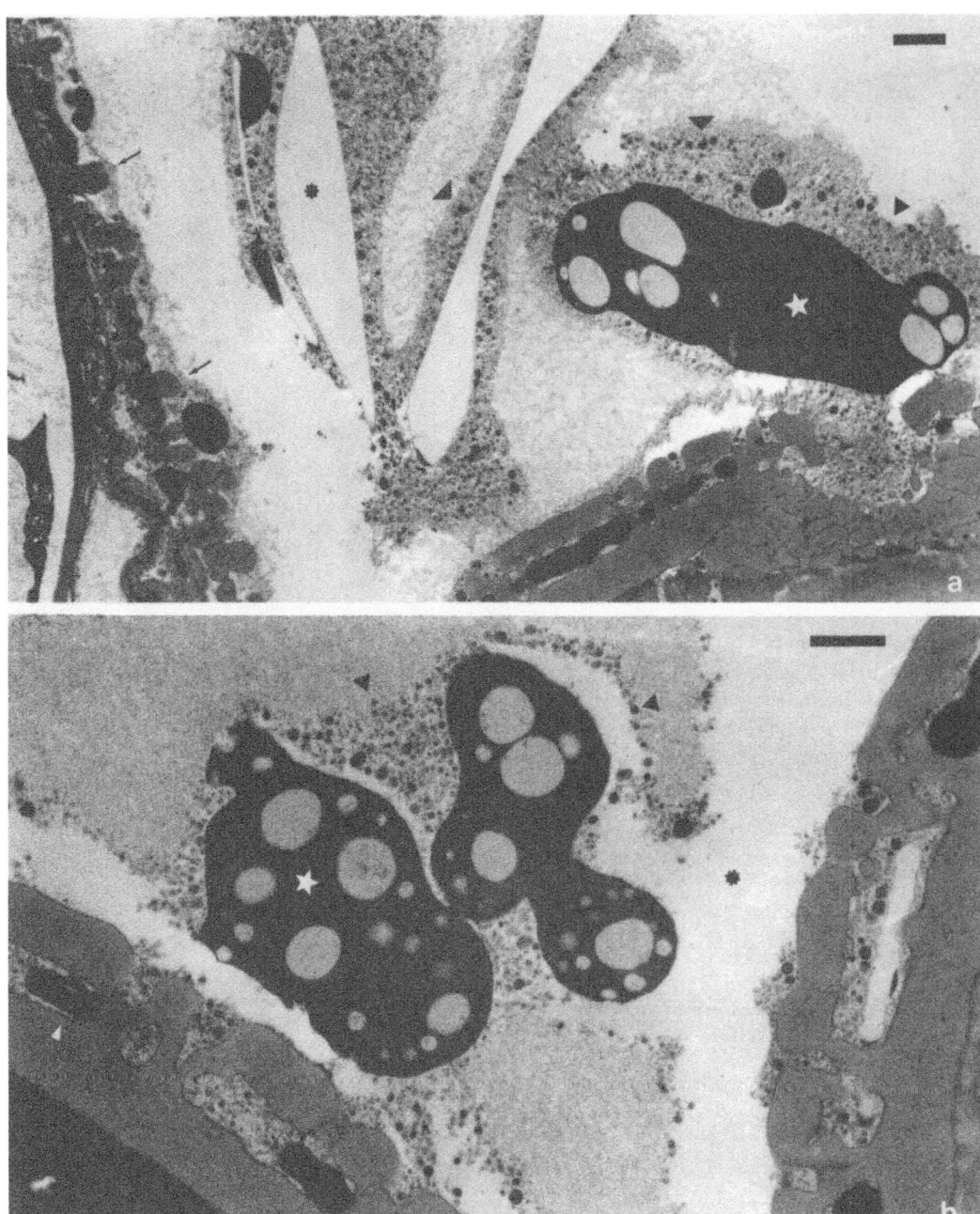

Fig. 8. *Tilia platyphyllos. a, b* after tapetum degeneration is completed, the pollenkitt is floating between mature pollen grains. Both pk components – embedded in a granular matrix – can still be distinguished as blistered lumps (white asterisks, formerly the non-plastidal, SER-generated component) and as abundant mostly small and electron-dense droplets (black arrowheads, resulting from the plastidal pk component). Compact or lamellated (white arrowhead) pollenkitt lumps and small droplets fill in part the intercolumellar spaces. Ubisch bodies (arrows) and pollenkitt droplets border the former loculus wall. Electron-transparent, "empty" spaces (black asterisks) between the floating pk components may either be due to some washing-out of unstable material during fixation or due to plasmolytic effects of fixation. – Bar: 1 μm

Discussion

Fixation quality. If discussing developmental aspects of cells with high metabolism the aspect of fixation quality should not be overlooked. The overall fixation quality of conventionally, i.e. of chemically fixed and prepared material has increased enormously in recent years, especially in highly metabolic plant cells. The fixation quality of these cells is now finally adequate to that of cells with reduced metabolism (cf. the generally accepted criteria for good fixations in HAYAT 1989). This should be borne in mind when considering the results of former tapetum investigations: the so-called "degeneration processes" of tapetal cells should be more precisely defined. Furthermore, some older observations concerning structure and development of organelles could be misinterpretations due to poor fixation quality, often causing fixation artifacts, and former results may therefore need some re-interpretations. It should be stressed that advanced fixation quality can hardly affect the features of dead cell inclusions like lipid droplets. This is demonstrated in two ways: (1) during early and medium stages of tapetum development (this is during early mid and late microspore stage) pollenkitt droplets in the plastids and within the cytoplasm widely retain their shape, dimension, and fine structure. This is in accordance with investigations comparing chemical fixations with physical methods (cryofixation as high pressure freezing, HESS & HESSE 1993), and this can even be seen in the studies of LOMBARDO & CARRARO (1976a, b) and COUSIN (1979), who supposed two sources of pk production in the tapetal cells (see also below). (2) At the end of the maturation the retention of pollenkitt lipids seen in TEM is often poor compared with living pollen seen in the light microscope. Only osmium tetroxide used alone as fixation medium preserves (all?) the otherwise widely extracted lipid material on and in the exine, while glutaraldehyde alone or followed by OsO_4 cannot stabilize lipid accumulations (ROWLEY & EL-GHAZALY 1992). But even without (any?) loss of lipids using this OsO_4-technique the pollenkitt looks extremely dense in TEM: no difference in structure is observed, only some lamellations depending on lipid self-assembly. The complexity of pk formation and composition can sufficiently be demonstrated at the CTEM level studying all developmental stages, if advanced chemical fixation protocols are used. Postfixation by Os–FeCN strongly improves the structural preservation of organelles in cells with extremely high metabolism like tapetal cells (WEBER 1992a).

Pollenkitt components. Already in publications from the 70' there was considerable evidence that more than a single tapetal organelle forms the pollenkitt components but because of the comparably poor fixation standard of this time there was no strict proof for it (LOMBARDO & CARRARO 1976a, b; COUSIN 1979; for review, e.g., ECHLIN 1971, and PACINI & al. 1985: 165 ff). It has recently become evident that pollenkitt can indeed be formed by various organelles at different times (TIWARI & GUNNING 1986, CHEN & al. 1988, POLOWICK & SAWHNEY 1990, EL GHAZALY & NILSSON 1991, MURGIA & al. 1991, CIAMPOLINI & al. 1993). Clear evidence that in *Apium nodiflorum* (1) aggregates accumulating in SER-vesicles, and (2) droplets from plastids form the pollenkitt was given by WEBER (1992b). In orchids, however, plastids do not contribute to the elastoviscin, the special pollenkitt of many *Orchidaceae* (WOLTER & al. 1988).

In contrast to my former view (HESSE 1978) in *Tilia platyphyllos* the pollenkitt-contributing lipids and/or carotenoids arise from two organelles at different times long before tapetum degeneration takes place: the SER-vesicles and the modified plastids (elaioplasts). The production of the first pk precursor type takes place in the cytoplasm starting as early as in the tetrad stage. The droplets become tightly enclosed by SER-profiles. This first type corresponds exactly to the non-plastidal, so-called "cytoplasmic" pk precursors in *Ledebouria* (HESS & HESSE 1993). But neither the improved chemical fixation and structure preservation by Os–FeCN (*Tilia*, see Results) nor the cryoimmobilization by high pressure freezing and cryo-substitution (*Ledebouria, Tillandsia,* HESS & HESSE 1993) allows unequivocally to decide whether the SER-enclosed lipid droplets accumulate within the lumen of SER-vesicles or external to these vesicles. In contrast WEBER (1992b) has shown that in *Apium nodiflorum* one of the two pollenkitt precursors accumulate within membrane-bound domains continuous with SER. The first pk type of *Tilia platyphyllos* is accumulated in the tapetal cytoplasma during exine formation and shortly before the first microspore mitosis. The production of the second pk component, which presumably contains carotenoids (DOBSON 1988, HESS & HESSE 1993) takes place – as shown before (HESSE 1978) – within a peculiar plastid type (elaioplasts) and starts shortly before the first haploid pollen mitosis. From the starting point of the production of both precursor types, but especially during the main secretion period, even in brightfield images, both types exhibit quite different electrondensity.

Analysis of pollenkitt composition by EFTEM. Pollenkitt of several taxa was shown as an inhomogeneous mixture of saturated and/or unsaturated lipids by thin-layer chromatography (DOBSON 1988). But at the CTEM level it is impossible to differentiate between various lipids, only analytical TEM can solve the problem. A further proof for pollenkitt diversity is based on the application of a recently developed type of EMs, the "filtering microscopes". These TEMs are equipped with an integrated electron spectroscopic filter (EFTEM = Energy Filtering TEM), which clearly allows to demonstrate by a specialized imaging technique that the pollenkitt precursors have different structure and varying density. Then even the seemingly homogeneous exine coating lipid droplets are structurally different.

The following passage should elucidate the EM technique applied. Electrons emitted from an electron gun can either be elastically or inelastically scattered by atoms of the specimen. The elastically scattered electrons form the familiar electron image associated with a conventional TEM, giving images only in the common brightfield or darkfield mode. The normally excluded inelastically scattered electrons can contribute additional information by their characteristic energy loss. Spectroscopic images (by ESI = Elemental Spectroscopic Imaging) can be used to improve specimen contrast by grey level discrimination and/or to detect chemical elements. If compared to conventional brightfield images even zero loss images of extremely thin unstained sections improve the contrast in seemingly homogeneous droplets. "Contrast tuning" means selecting that part of the Electron Energy Loss Spectrum (EELS), where the contrast is optimized (REIMER & al. 1991; FROMM & al. 1992; FEHRENBACH & al. 1991, 1992): i.e. viewing and comparing the very same sample in the zero loss, in plasmon loss and in electron energy loss modes, respectively. This techniques reveal normally hidden differences in lipid electron density.

While an EFTEM so far cannot determine the chemical composition of lipids and/or carotenoids, it helps to detect the otherwise hidden structural inhomogenity of droplets. Contrast tuning gives a better grey level discrimination also in lipid and/or carotenoid droplets and a structure-specific contrast can be observed (FEHRENBACH & al. 1991, 1992). Sufficient grey level discrimination is provided by optimized contrast from variations in electron density of lipid droplets, thereby permitting images of the different components. From this a heterogeneous chemical composition of seemingly homogeneous pollenkitt components can be concluded (HESSE 1991).

Conclusions

It is now clear that at least in some taxa more than a single tapetal cell organelle contributes to pollenkitt formation, i.e. the plastids and the SER connected lipid droplets. From the beginning of pollenkitt production onwards both pk components are by no means structurally homogeneous, although at maturity the mostly highly electron-dense pollenkitt mixture often looks very compact. Before and after their release from the organelle where they are generated all the pk components appear in CTEM and especially in an EFTEM as widely inhomogeneous. In late developmental stages – during and after tapetum degeneration – both pollenkitt precursors evidently fuse forming the proper, predominantly electron-dense, generally homogeneous pollenkitt (REZNICKOVA & DICKINSON 1982, PACINI 1990, ROWLEY & EL-GHAZALY 1992) and is transferred to the mature pollen (WEBER 1991).

The author is highly indebted to Dr MARTINA WEBER for manifold collaboration, he thanks also MTA CHRISTA GRUBMANN, Dr MICHAEL W. HESS, and Dr HEIDEMARIE HALBRITTER for technical assistance. He is especially grateful to Prof. Dr WALTRAUD KLEPAL (Institute of Zoology, University of Vienna) for the use of the ZEISS EM 902, and to Mag. MICHAEL G. SCHLAG for improving the English.

References

CHEN, Z., WANG, F., ZHOU, F., 1988: On the origin, development and ultrastructure of the orbicules and pollenkitt. – Grana **27**: 273–282.

CIAMPOLINI, F., NEPI, M., PACINI, E., 1993: Tapetum development in *Cucurbita pepo* (*Cucurbitaceae*). – Pl. Syst. Evol. [Suppl.] **7**: 13–22.

COUSIN, M.-Th., 1979: Tapetum and pollen grains of *Vinca rosea* (*Apocynaceae*). – Grana **18**: 115–128.

DOBSON, H. E. M., 1988: Survey of pollen and pollenkitt lipids – chemical cues to flower visitors? – Amer. J. Bot. **75**: 170–182.

ECHLIN, P., 1971: The role of the tapetum during microsporogenesis of angiosperms. – In HESLOP-HARRISON, J., (Ed.): Pollen. Development and physiology, pp. 41–61. – London: Butterworths.

EL-GHAZALY, G. A., NILSSON, S., 1991: Development of tapetum and orbicules of *Catharanthus roseus* (*Apocynaceae*). – In BLACKMORE, S., BARNES, S. H., (Eds): Pollen and spores. Patterns of diversification, pp. 317–330. – Syst. Assoc. Vol. **44**. – Oxford: Clarendon Press.

FEHRENBACH, H., RICHTER, J., SCHNABEL, P. A., 1991: Improved preservation of phospholipid-rich multilamellar bodies in conventionally embedded mammalian lung tissue – an electron spectroscopic study. – J. Microscopy **162**: 91–104.

– – – 1992: Electron spectroscopic study (ESI, EELS) of Nanoplast-embedded mammalian lung. – J. Microscopy **166**: 401–416.

FROMM, I., REIMER, L., RENNEKAMP, R., 1992: Investigation and use of plasmon losses in energy-filtering transmission electron microscopy. – J. Microscopy **166**: 257–272.

HAYAT, M. A., 1989: Principles and techniques of electron microscopy. 3rd edn. – London: MacMillan.

HENNING, J. A., TEUBER, L. R., 1992: Identification of pollenkitt variation among Alfalfa germplasm sources. – Crop Sci. **32**: 653–656.

HESS, M. W., HESSE, M., 1993: Ultrastructural observations on tapetum development in freeze fixed *Ledebouria socialis* (*Hyacinthaceae*). – Planta (in press).

HESSE, M., 1978: Entwicklungsgeschichte und Ultrastruktur des Pollenkitts bei *Tilia* (*Tiliaceae*). – Pl. Syst. Evol. **129**: 13–30.

– 1991: Different mass densities by ESI in pollenkitt and pollen wall layers. – European J. Cell Biol. **55**, [Suppl.] **34**: 11.

KEIJZER, C. J., 1987: The process of anther dehiscence and pollen dispersal. II. The formation and the transfer mechanism of pollenkitt, cell-wall development of the loculus tissues and a function of orbicules in pollen dispersal. – New Phytol. **105**: 499–507.

LOMBARDO, G., CARRARO, L., 1967a: Tapetal ultrastructural changes during pollen development. I. Studies on *Anthirrhinum maius*. – Caryologia **29**: 113–125.

– CARRARO, L., 1976b: Tapetal ultrastructural changes during pollen development. III. Studies on *Gentiana acaulis*. – Caryologia **29**: 345–349.

MURGIA, M., CHARZYNSKA, M., ROUGIER, M., CRESTI, M., 1991: Secretory tapetum of *Brassica oleracea* L.: polarity and ultrastructural features. – Sex. Pl. Reprod. **4**: 28–35.

PACINI, E., 1990: Tapetum and microspore function. – In BLACKMORE, S., BARNES, S. H., (Eds): Microspores: evolution and ontogeny, pp. 213–237. – London: Academic Press.

– FRANCHI, G. G., HESSE, M., 1985: The tapetum: its form, function, and possible phylogeny in *Embryophyta*. – Pl. Syst. Evol. **149**: 155–185.

POLOWICK, P. L., SAWHNEY, V. K., 1990: Microsporogenesis in a normal line and in the *ogu* cytoplasmic male-sterile line of *Brassica napus*. – Sex. Pl. Reprod. **3**: 263–276.

REIMER, L., RENNEKAMP, R., FROMM, I., LANGENFELD, M., 1991: Contrast in the electron spectroscopic mode of a TEM. IV. Thick specimen imaged by the most probably energy loss. – J. Microscopy **162**: 3–14.

REZNICKOVA, S. A., DICKINSON, H. G., 1982: Ultrastructural aspects of storage lipid mobilization in the tapetum of *Lilium hybrida* var. enchantment. – Planta **155**: 400–408.

ROWLEY, J. R., EL-GHAZALY, G., 1992: Lipid in wall and cytoplasm of *Solidago* pollen. – Grana **31**: 273–283.

TIWARI, S. C., GUNNING, B. E. S., 1986: An ultrastructural, cytochemical and immuno-fluorescence study of postmeiotic development of plasmodial tapetum in *Tradescantia virginiana* L. and its relevance to the pathway of sporopollenin secretion. – Protoplasma **133**: 100–114.

WEBER, M., 1991: The transfer of pollenkitt in *Smyrnium perfoliatum* (*Apiaceae*). – Ann. Bot. **68**: 63–68.

– 1992a: Nature and distribution of the exine-held material in mature pollen grains of *Apium nodiflorum* L. (*Apiaceae*). – Grana **31**: 17–24.

– 1992b: The formation of pollenkitt in *Apium nodiflorum* (*Apiaceae*). – Ann. Bot. **70**: 573–577.

WOLTER, M., SEUFERT, C., SCHILL, R., 1988: The ontogeny of pollinia and elastoviscin in the anther of *Doritis pulcherrima* (*Orchidaceae*). – Nordic J. Bot. **8**: 77–88.

Address of the author: MICHAEL HESSE, Institute of Botany and Botanical Garden of the University of Vienna, Rennweg 14, A-1030 Wien, Austria.

Pl. Syst. Evol. [Suppl.] 7: 53–62 (1993)

Secretory events in the freeze-substituted tapetum of the orchid *Pterostylis concinna*

M. A. Fitzgerald, D. M. Calder, and R. B. Knox

Key words: *Orchidaceae, Pterostylis concinna.* – Tapetum, freeze-substitution, exine formation, pollencoat.

Abstract: Three distinct secretory events in tapetal cells correspond to key processes of microspore development: microspore release and expansion, exine development, and pollencoat formation and dispersal. First, at the tetrad stage, the lower part of the radial and the inner tangential tapetal walls contain protein and are markedly thick and fibrillar. The subsequent absence of this extra wall after microspore release indicates that it may contain 1,3-β glucanase, active in callose dissolution. Using freeze-substitution to investigate tapetal development demonstrates that the secretory tapetal cell of *Pterostylis concinna* is functional until exine formation. Tapetal cell material then reorganizes for two further secretory events: second, the exine of the young microspore is developed, tapetal cells lose their characteristic shape, the plasma membrane is no longer detectable, and the tapetal contents flow out into the locule. Third and final tapetal release is electron-opaque pollen-coat material occurring before microspore mitosis.

There are two types of tapeta: secretory and amoeboid (Pacini 1990). The transfer of products from the secretory tapetum is an integral part of pollen grain development. Products secreted from the tapetum are implicated in microspore release, microspore growth, and exine development. Tapetal cells must be actively synthesising and secreting substances such as 1,3-β glucanase for microspore release, material for exine growth and pollen glue. These activities require them to be functional cells with a plasma membrane able to regulate passage of material. Eventually tapetal cells die and the plasma membrane becomes porous and ineffective. At which point in microspore development does the tapetum cease to be a functional cell, i.e., when does the plasma membrane lose its integrity?

The state of the tapetal plasma membrane can be assessed using transmission electron microscopy (TEM). There are several problems inherent with conventional TEM investigation, primarily, separating life state conditions from fixation artefacts. Plasma membranes are particularly susceptible to artefacts of chemical fixation because their proteins are often the first in the cell to contact fixative. During the several minutes needed to fix cytoplasmic contents, osmotic adjustments due to the fixative place stress on the dead plasma membrane as liquids pass in or out of the cytoplasm. The fixed membrane does not have the same flexibility as the living fluid membrane to regulate passage through it, often resulting in undulation and whorling of the plasma membrane (KELLENBERGER 1991). These effects are comparable to natural cell senescence, making it difficult to discriminate between fixation artefacts and actual senescence. Freeze-substitution, as a method of fixation for biological TEM-investigations, and in particular, for membrane preservation (HYDE & al. 1991) has gained wide acceptance, and has changed at least in part, some our views on membrane formation and function. With this method, cells are rapidly frozen and are maintained frozen, while fixation chemicals are introduced. After fixation, cells are slowly warmed to room temperature. Hence, images should represent life-state-conditions as at time of freezing. When freezing anthers, both microspores and tapetum are fixed. Actively developing microspores should be healthy cells suitable as a control for fixation quality. We should assume that, if these are well fixed, then tapetal cells should also be. Therefore quality of fixation can be separated from the life state of tapetal cells. The present study investigates secretory tapetal activity and its role in pollen grain development in the orchid *Pterostylis concinna*.

Material and methods

Pterostylis concinna FITZG (*Orchidaceae*) flowers were collected at different stages of development from a site at Brisbane Ranges National Park in the Shire of Bacchus Marsh, Victoria, Australia. Anthers were removed from the buds, sliced into thin pieces and immediately plunged into liquid propane, cooled to the temperature of liquid nitrogen. This process took less than 10 sec. The frozen pieces were transferred to vials precooled with liquid nitrogen and containing 2.5% OsO_4 in dry acetone and molecular sieve. The vials were sealed and placed at $-70°$C. After 3 days the vials were warmed as follows: $-20°$C for 2 h, 4°C for 1 h, and room temperature for 1 h. Specimens were placed in new vials and underwent 3×5 min changes of fresh, dry acetone. They were then infiltrated with Spurr's resin (for TEM) or LR White resin (for LM), reaching 100% resin after 2 days. Resin was changed daily for another 3 days, then the specimens were embedded.

Sections (60–90 nm) were cut on a Reichert Ultracut E ultramicrotome, using a diamond knife, were then collected on 50 mesh copper grids coated with formvar, stained in uranyl acetate (20 min), washed, stained with lead citrate (10 min), washed again. Sections were examined on a Jeol 1200EX transmission electron microscope (TEM) operating at 80 kV and a Siemens Elmiskop TEM operating at 60 kV. LR White resin sections (1 μm) were cut on the same microtome, stained in 1% amido black in 7% acetic acid for proteins (10 min), then rinsed in 1% acetic acid in distilled water. They were examined on an Olympus CH light microscope using bright field.

Results

Changing morphology of the tapetum and its interactions with pollen grain ontogeny is presented for a developmental sequence in the anther spanning the period from meiotic tetrads surrounded by callose, until microspore mitosis.

Tetrad stage. The tapetum is closely appressed to tetrads which are surrounded by the original microsporocyte pectocellulosic wall, then by a callose wall (Figs. 1a, 2a, b). Tapetal cells are uninucleate. A conspicuous and unusual feature is the markedly thickened and fibrillar wall visible on the lower (proximal) portion of the tangential and the inner radial walls. Both wall parts, directly abutting tetrads, are amido black positive indicating the presence of protein (Fig. 1b). The outer

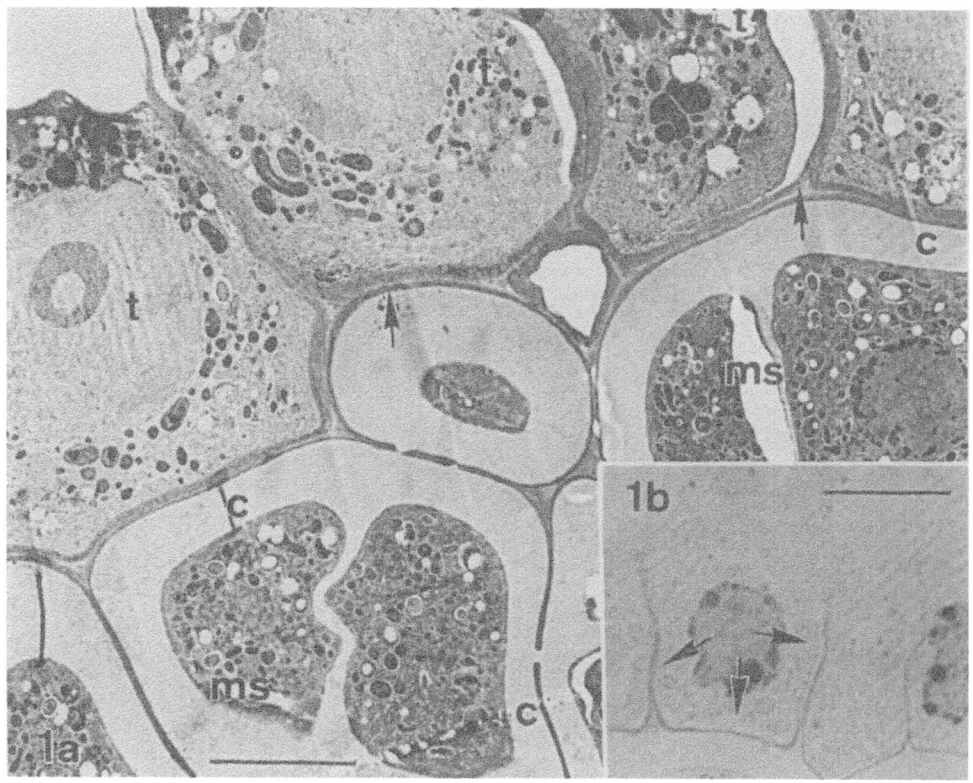

Fig. 1. *Pterostylis concinna.* Early microspore stage with callose surrounding microspores. *a* Microspores (ms) surrounded by callose wall (c), and the locule surrounded by a single layer of uninucleate tapetal cells (t) with thick lower radial and inner tangential walls (arrows). Bar: 5 μm. *b* Lower radial and inner tangential tapetal walls stain positively with amido black (arrows). Bar: 10 μm

tangential walls and upper (distal) parts of the inner radial walls are thin and not fibrillar. Figure 2b shows unusual electron opaque deposits adherent to the plasma membrane. In the tapetal cells slightly condensed Golgi bodies, mitochondria, ER, and plastids dominate.

Free microspore stage. Following microspore release, the tapetal cell walls have lost their inner, thick, proteinaceous layer (Fig. 3a). During the free microspore stage, tapetal cells synthesize two distinct types of material (Fig. 3c), one electron opaque and the other partially electron opaque with similar appearance to the exine. These two materials fill the tapetal cell cytosol. Until their synthesis is complete the microspore foot layer remains very thin, although bacula and intine develop. Figure 3 shows the sequence of reorganization of tapetal cell contents to form the two types of material. Osmiophilic droplets and organelles containing profiles of membranes begin to form, but cell shape and plasma membrane are still intact (Fig. 3a). Osmiophilic droplets in Fig. 3b are larger and more numerous, organelles are becoming less identifiable, but the plasma membrane and cell shape are still unchanged. After this the tapetal cell is beginning to lose its characteristic shape,

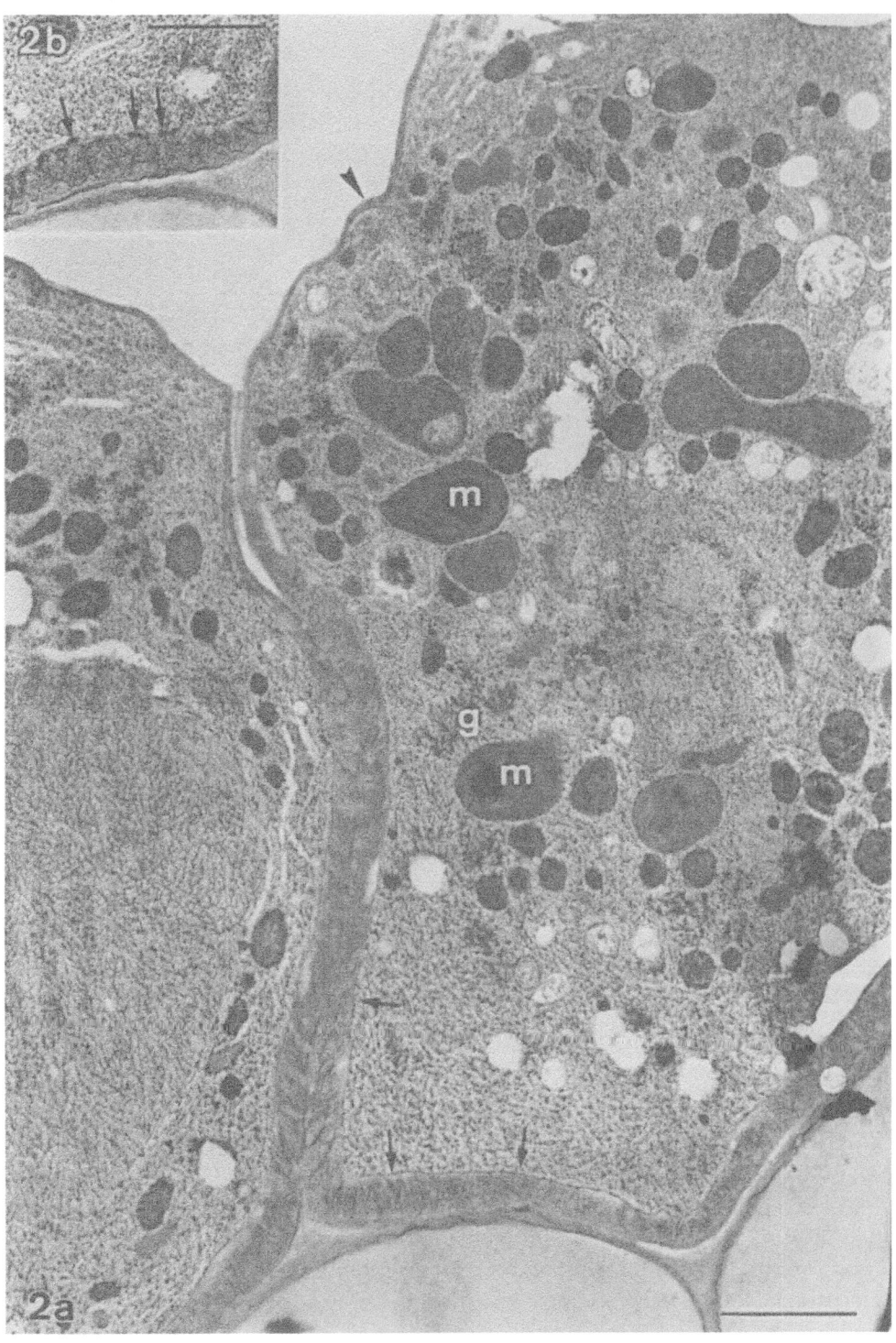

Fig. 2. *Pterostylis concinna*. Tapetum cells at tetrad stage. *a* Lower radial and inner tangential tapetal walls are thick and fibrillar (arrows). Tapetum contains numerous Golgi bodies (g) and mitochondria (m). Bar: 1 μm. *b* Electron opaque globules beneath plasma membrane (arrows). Bar: 0.5 μm

and its contents have condensed into the previously described electron opaque and partially electron opaque material (Fig. 3c). Figure 3d shows the same stage as 3c, and microspore intine and bacula are well developed, but the foot layer remains thin. In the last stage the tapetal material is now present in the anther locule and the tapetal plasma membrane is no longer visible (Fig. 3e). Low electron opaque material inks microspore exine elements and tapetum. Following this secretory event, the microspore exine and the foot layer are now well developed (Fig. 4). The tapetal cell contents then detach from the anther wall and all the remaining electron opaque material settles among bacula, and links adjacent microspores. Nevertheless radial and outer tangential tapetal walls are still identifiable.

Microspore mitosis. Following microspore mitosis the tapetal wall remnants still adhere to endothecial cells. Tapetal material forms a conspicuous pollencoat on the surface and among the bacula, an appearance similar to that of the microspore stage (Fig. 5).

Discussion

Use of freeze-substitution. The use of freeze-substitution has given a new understanding of developmental processes at the sub-cellular level. In the past it has been difficult to draw conclusions about cell physiology using TEM, because the time needed for fixatives to penetrate and kill cells affects cell physiology to a degree detectable at the resolution of the TEM. Using freeze-substitution in investigations of other physiological processes has produced results that conflict with what were previously accepted facts. For example, HYDE & al. (1991) showed that formation of the cell plate in zoospores undergoing mitosis occurs from positioning of membrane sheets (detected by freeze-substitution) rather than fusion of vesicles containing wall material (previously detected after chemical fixation). This finding has revolutionised accepted theories of the mitotic process. CANNY & MCCULLY (1986) used freeze-substitution to show that transpirational water travels symplastically from leaf vein endings to stomata. Studies on water and solute pathways in mangrove leaves by FITZGERALD & ALLAWAY (1991) and on salt secretion in mangroves (FITZGERALD & al. 1992) have shown that salt actually travels symplastically with the transpiration water to the salt glands where it is secreted, and not apoplastically and separate from the water, as previously thought. These examples show that freeze-substitution is an effective method for investigation of physiological processes.

Secretory events in the tapetum. Physiological interactions between tapetum and microspores involve synthesis within the tapetum of material used for microspore release, exine formation, and pollen grain cohesion. This material is transferred from tapetum to microspores in three distinct secretory events.

First secretory event, microspore release. In *Pterostylis*, a unique protein-aceous layer is deposited inside the inner tapetal cell walls that are in direct contact with the tetrads. It persists until microspore release, when the tapetal wall becomes a uniformly thin primary wall around the entire cell perimeter. In other taxa, 1,3 β-glucanase is a tapetally derived enzyme active in callose dissolution (STIEGLITZ & STERN 1973, SCOTT & al. 1991). It is intriguing that following microspore release, the proteinaceous layer abutting lower radial (i.e. proximally

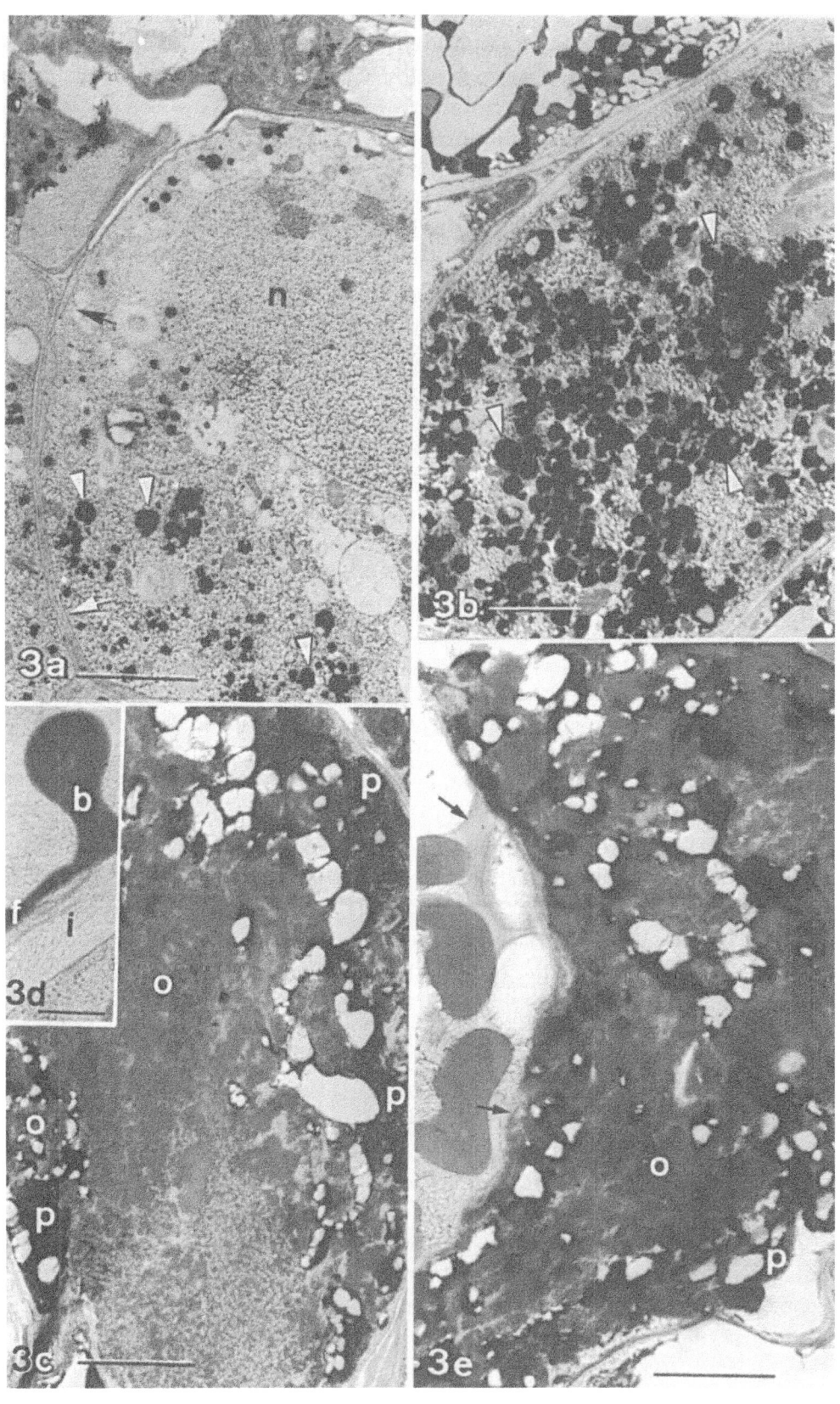

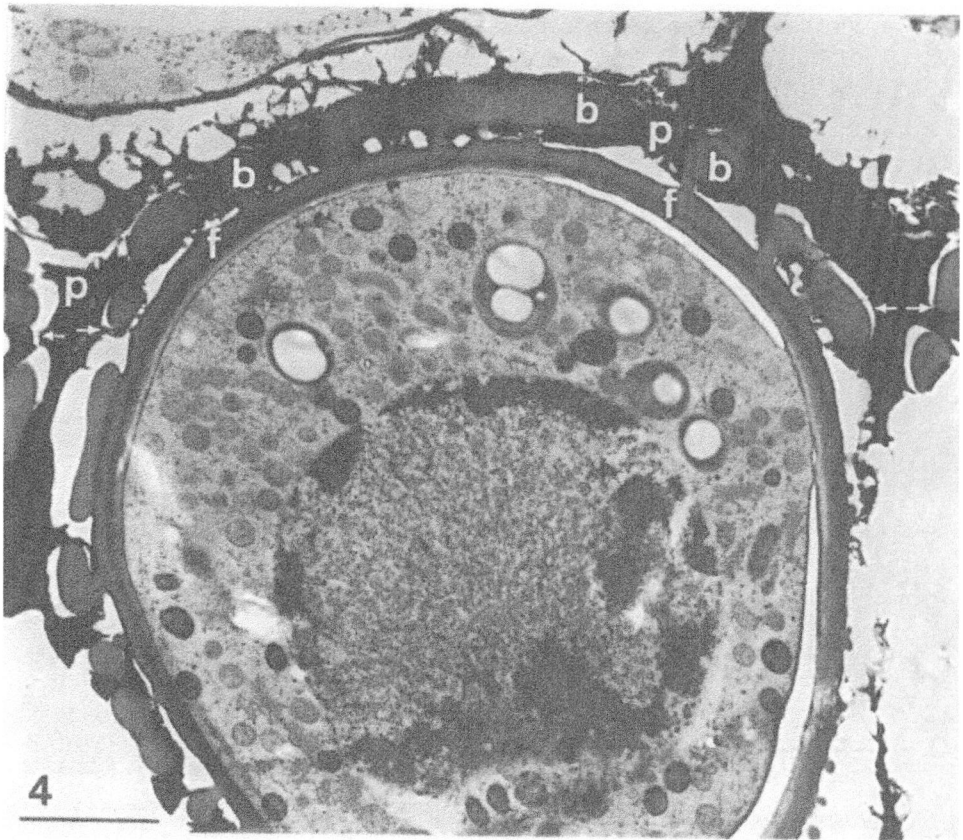

Fig. 4. *Pterostylis concinna*, late microspore stage. The exine is completed, note thick foot layer (f), and the third and final tapetal release has occurred. Prior to microspore mitosis, electron opaque material (p) has settled among bacula (b), and links adjacent microspores (arrows). Very little tapetal material adheres to endothecial cells. Bar: 3 μm

oriented parts) and inner tangential tapetal walls is no longer evident, perhaps indicating that the protein observed might be stored 1,3-β glucanase.

Following microspore release, the anther locule is filled with electron opaque material, probably a mixture of polysaccharides derived from callose and cellulose dissolution, suspended in anther locular fluid. These polysaccharides provide sugars

Fig. 3. Sequence of tapetal reorganisation in *Pterostylis concinna* during free microspore stage. *a* Plasma membrane is intact and appressed to tapetal walls which are uniformly thin following microspore release (arrows). Nucleus (n) and cell shape intact. Electron opaque globules are forming (arrowheads). Bar: 1.5 μm. *b* Electron opaque globules are larger and more numerous (arrowheads). Cell shape still intact. Bar: 1 μm. *c* Tapetal material is reorganised into two types; one electron opaque (o) and the other partially electron opaque (p). Bar: 1 μm. *d* Same as stage *c*, the pollen wall foot layer is thin (f), but intine (i) and bacula (b) are prominent. Bar: 1.5 μm. *e* Cell shape is lost and the partially electron opaque material (p) starts to enter the anther locule (arrows), while the opaque material (o) is left behind. Bar: 1.5 μm

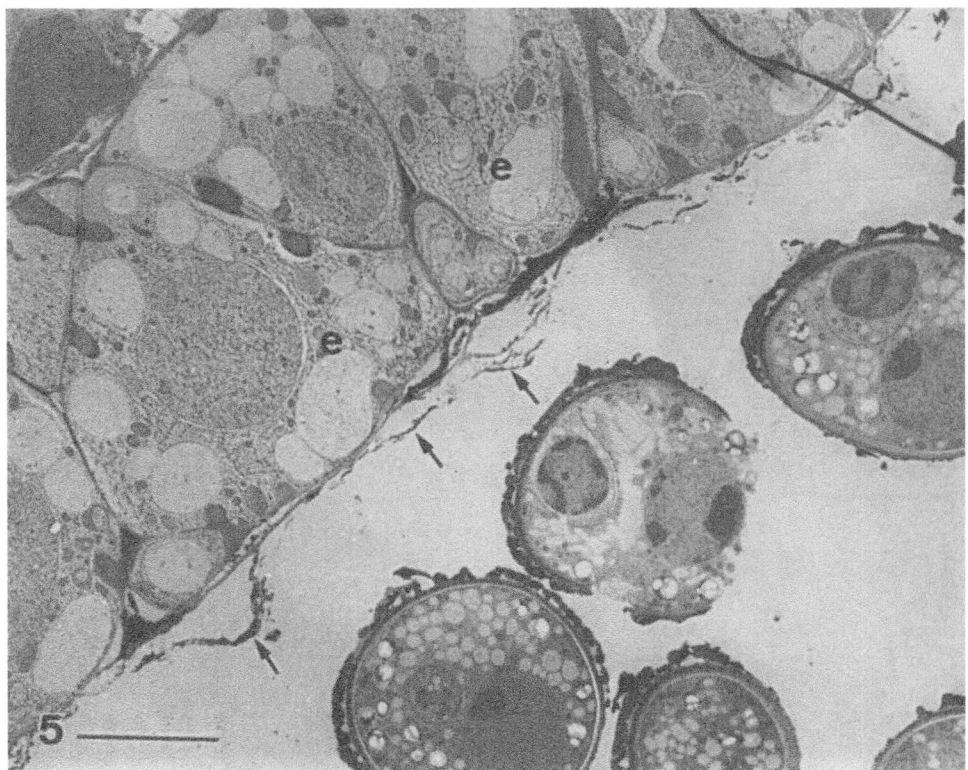

Fig. 5. *Pterostylis concinna*, young pollen grain stage. Tapetal material is still among baculae following mitosis, and, as in Fig. 4, very little tapetal material (arrows) adheres to endothecial cells (e). Bar: 10 µm

for microspore growth. The microspore volume increases 2.5 fold reaching full size before mitosis, and the footlayer remains thin (FITZGERALD & al., in prep.). The radial tapetal walls become convoluted.

During microspore growth, major changes occur in the tapetum after which only two types of tapetal material are recognizable: one very electron opaque, and one partially electron opaque similar to the exine material. The plasma membrane is still intact and the cell is still functional.

Second secretory event, exine formation. The second secretory event is involved in exine formation. Just prior to the tapetal secretion, the microspore footlayer is still thin. Presumably it remains thin during microspore growth to allow expansion. The partially electron opaque material is secreted into the anther locule when the microspores reach their full size. The microspore exine is then fully developed, including an expanded footlayer. Only the electron opaque material remains in the tapetum.

Third secretory event, cohesion of pollinia. The third and final tapetal release occurs after exine development and prior to microspore mitosis. The tapetal cell plasma membrane is not identifiable at this stage, indicating that it is porous and ineffective. All remaining tapetal material has condensed into an osmiophilic, lipidic substance which is released prior to microspore mitosis and settles among bacula, acting as a pollencoat, linking adjacent pollen grains and persisting through the rest of the developmental process. In *Pterostylis concinna*, there are no exine bridges between the pollen grains, so that the lipid mixture from final tapetal release is presumably pollen glue to bind the pollen grains into a pollinium. The final material does neither resemble exine material nor is it connected to the exine, hence it is not viscin (HESSE 1984). Neither is it elastoviscin, as special tapetal cells forming this substance were not observed (SCHILL & WOLTER 1985). Following the third tapetal release, only the outer distal tangential tapetal walls remain, hence leaving no lipid droplets from which pollenkitt form could (HESSE 1979, PACINI & CASADORO 1981). In *Pterostylis concinna*, glue material is necessary to hold pollen grains in a pollinium, but in this orchid species the glue type does not fall under previously defined pollencoat types (KNOX 1984).

Conclusions

Using freeze-substitution to investigate tapetal processes has demonstrated that the secretory tapetal cell of *Pterostylis concinna* is functional until after the second secretory event and subsequent exine formation. Enzymes involved in microspore release may be actively secreted from external tapetal wall thickenings, exine forming materials are secreted following microspore expansion, then the tapetal cell membrane degenerates. Pollen glue is released after complete tapetal breakdown, which is markedly earlier than in any other species previously reported, occurring before microspore mitosis. In most other species the preformed pollencoat is deposited just prior to anther dehiscence (KNOX 1984, PACINI 1990). This pollen glue does not fit previously defined types, hence may be another evolutionary strategy towards entomophily.

Part of this work was done with the help of Dr STEVE BLACKMORE and Ms SUE BARNES at The Natural History Museum, Cromwell Road, London, and we thank them for providing the facilities and for their assistance. We also thank the Australian Research Council for financial assistance.

References

CANNY, M. J., MCCULLY, M. E., 1986: Locating water soluble vital stains in plant tissues by freeze-substitution and resin embedding. – New Phytol. **115**: 511–516.

FITZGERALD, M. A., ALLAWAY, W. G., 1991: Apoplastic and symplastic pathways in the leaf of the grey mangrove *Avicennia marina* (FORSK.) VIERH. – New Phytol. **119**: 217–226.

FITZGERALD, M. A., ORLOVICH, D. A., ALLAWAY, W. G., 1992: Evidence that abaxial leaf glands are the sites of salt secretion in leaves of the mangrove *Avicennia marina* (FORSK.) VIERH. – New Phytol. **120**: 1–7.

HESSE, M., 1979: Entwicklungsgeschichte und Ultrastruktur von Pollenkitt und Exine bei nahe verwandten entomo- und anemophilen Angiospermen: *Salicaceae, Tiliaceae und Ericaceae*. – Flora **168**: 540–557.

– 1984: An exine architecture model for viscin threads. – Grana **73**: 69–75.

HYDE, G. J., LANCELLE, S., HEPLER, P. K., HARDHAM, A. R., 1991: Freeze-substitution reveals a new model for sporangial cleavage in *Phytophthora*, a result with implications for cytokinesis in other eukaryotes. – J. Cell Sci. **100**: 735–746.

KELLENBERGER, E., 1991: The potential for cryofixation and freeze substitution: observations and theoretical considerations. – J. Microscopy **161**: 183–203.

KNOX, R. B., 1984: The pollen grain. – In JOHRI, B. M., (Ed.): Embryology of angiosperms, pp. 197–271. – Berlin: Springer.

PACINI, E., 1990: Tapetum and microspore function. – In BLACKMORE, S., KNOX, R. B., (Eds): Microspores: Evolution and Ontogeny, pp. 213–237. – London: Academic Press.

– CASADORO, G., 1981: Tapetum plastids of *Olea europaea*. – Protoplasma **106**: 289–297.

SCHILL, R., WOLTER, M., 1985: Ontogeny of elastoviscin in the *Orchidaceae*. – Nordic J. Botany **5**: 575–580.

SCOTT, R., HODGE, R., PAUL, W., DRAPER, J., 1991: The molecular biology of anther differentiation. – Pl. Sci. **80**: 167–191.

STIEGLITZ, H., STERN, H., 1973: Regulation of 1,3-β glucanase activity in developing anthers of *Lilium*. – Develop. Biol. **34**: 169–173.

Address of the authors: M. A. FITZGERALD, D. M. CALDER and R. B. KNOX, School of Botany, The University of Melbourne, Parkville 3052, Victoria, Australia.

Pl. Syst. Evol. [Suppl.] 7: 63–74 (1993)

Cytochemical and ultrastructural evolution of orbicules in *Lilium*

C. CLÉMENT and J. C. AUDRAN

Key words: *Liliaceae, Lilium.* – Orbicule, ultrastructure, cytochemistry, cytomembrane, polygonal meshes network, sporopollenin architecture and biosynthesis.

Abstract: The Ubisch body of *Lilium* is investigated ultrastructurally and cytochemically. It consists of four concentric distinct zones: (1) the **orbicular core** surrounded by (2) a **cytomembrane** supporting a glycocalyx on which (3) the sporopollenin **orbicular wall** is built bordered by (4) a **peripheral sheet**. The latter covers the whole orbicular surface and is organized in a polygonal network. The orbicular sporopollenin is then supposed to be formed on a glycoprotein polygonal frame, accumulated on the cytomembrane surrounding the orbicular core and filled with lipids and polyphenolic compounds.

At the interface between diploid sporophytic tissues of the anther wall and haploid gametophytic cells, orbicules or Ubisch bodies are localized against tapetal cells (BHANDARI 1984). In *Spermatophyta*, they have been described only in anthers with secretory tapetum (PACINI & FRANCHI 1991). Many functions have been attributed to orbicules but none is clearly demonstrated. Many authors think that orbicules represent a transient reserve of sporopollenin for pollen exine (HESLOP-HARRISON & DICKINSON 1969, RISUEÑO & al. 1969, BANERJEE & BARGHOORN 1971, BHANDARI 1984). They should be involved in pollen dispersal from the locule, acting altogether as a non-wettable surface so that pollen is more easily detached (KEIJZER 1986, RAJ & EL-GHAZALY 1987).

Orbicular ontogenesis often starts during meiosis. EL GHAZALY & JENSEN (1986) have reported that proorbicule initiation takes place from lipid globules synthesized in the tapetal cytoplasm. Each proorbicule is covered by sporopollenin concurrently with pollen exine synthesis. The orbicular wall then undergoes a gentle retraction which leads to a diminution of the orbicule (REZNICKOVA & WILLEMSE 1980). The orbicular wall is presumed to be synthesized on a glycocalyx supported by a membrane-like lamellae that surrounds the proorbicule (ROWLEY & SKVARLA 1974, REZNICKOVA & WILLEMSE 1980).

Because of their small size (rarely exceeding 5 μm) the Ubisch bodies have usually been studied at the ultrastructural level. Especially the orbicular wall is compared to the pollen exine. Both give positive reactions for proteins, acid polysaccharides, and unsaturated lipids (EL-GHAZALY & JENSEN 1987).

Less information is available concerning the cytochemistry and ultrastructure of the whole orbicules during its development (e.g., REZNICKOVA & al. 1980, CLÉMENT & AUDRAN 1992). In the present paper we investigated cytochemically how orbicular wall sporopollenin is formed.

Material and methods

Anthers of *Lilium* cv. enchantment were excised from flower buds at three different stages of pollen development in order to get stages comparable with previous works (REZNICKOVA & WILLEMSE 1980, EL-GHAZALY & JENSEN 1986): (1) **young microspore stage** (deposition of orbicular sporopollenin), (2) **vacuolate microspore stage** (Ubisch bodies reach their maximum size), (3) **young bicellular pollen grain stage** (maturation of orbicules).

Polarizing microscope (PM). Anthers were fixed in FAA (4h), dehydrated in an alcohol series, and embedded in paraffin. Sections of 6 μm were deparaffined in xylene, hydrated in decreasing alcohols series and observed in water under a polarizing microscope.

Transmission electron microscope (TEM). Anthers are treated following various protocols: – GOBA: fixation (4h) with 4% glutaraldehyde in 0,1 M cacodylate buffer at pH 7,5 in presence of 1% Alcian blue 8 GX, and postfixed (4h) with 1% OsO_4 in veronal acetate buffer at pH 7,2. Some anthers were acetolysed according to ERDTMAN (1960) before the GOBA fixation. – GOR: fixation (4h) with 4% glutaraldehyde in 0,1 M cacodylate buffer at pH 7,5 and postfixation (4h) with 1% OsO_4 in veronal acetate buffer at pH 7,2. Fixative and post-fixative contained 0,3% of Ruthenium red to preserve pectins (ROLAND 1978). – GFH: fixation (4h) with 4% glutaraldehyde in 0,1 M cacodylate buffer at pH 7,5. After washing, the Ferric Hydroxylamin reaction (ROLAND 1978) was applied to detect esterified pectins. – OsO_4: fixation (4h) in 1% OsO_4 in veronal acetate buffer at pH 7,2. – $KMnO_4$: fixation (2h) with 4% $KMnO_4$ and postfixation (2h) with 2% $KMnO_4$.

Anthers are then rinsed, dehydrated in an alcohol series and embedded in Spurr's resin. Thin sections treated according to the GOR, GFH, OsO_4 and $KMnO_4$ protocols were observed without staining. Concerning GOBA protocol treated material, many cytochemical tests were applied to detect neutral polysaccharides (Periodic Acid Thiosemicarbazide Silver Proteinate = PATAg: THIÉRY 1967, preceded or not by a H_2O_2 treatment), acid poly-saccharides (Phospho Tungstic Acid [PTA] low pH: THIÉRY & RAMBOURG 1974; PTA chromic acid: RAMBOURG 1969), proteins (PTA acetone: BENEDETTI & BERTOLINI 1963), unsaturated lipids (ThioSemiCarbazide Silver Proteinate = TSC SP: SELIGMAN & al. 1966) and polyphenolic compounds (Silver Proteinate = SP: MARINOZZI & al. 1977).

Additionally, the extraction of proteins has been realised by pronase treatment: ultrathin sections were immersed in 0.5% pronase solution in citrate buffer at pH 4.2 during 24 h at room temperature. After rinsing, the sections were tested with PTA acetone to detect residual proteins.

Results

The Ubisch bodies of *Lilium* are composed of four distinctive zones (Fig. 1): the **orbicular core** representing remnants of the proorbicule, an individualized **interface** between orbicular core and wall, the **orbicular wall** with bulges, and a **peripheral sheet** around the orbicular wall. Each zone has a constant dimension during development, except the orbicular wall, which increases between stage 1 and 2 and then decreases during maturation (Table 1).

The four zones have been characterized ultrastructurally and cytochemically (Table 1).

Table 1. Reactivity and rise of the different zones of Ubisch body to TEM cytochemical tests

Stages		Orbicular core			Orbicular core-wall interface			Orbicular wall			Orbicular peripheral sheet		
		1	2	3	1	2	3	1	2	3	1	2	3
Proteins:	PTA acetone	+	+	+	+	+	+	+	+	+	+	+	+
	Pronase + PTA acetone	−	−	−	−	−	−	−	−	−	+	+	+
Lipids:	Total OsO$_4$	+	+	+	+	+	+	+	+	+	+	+	+
	Unsaturated TSC-SP	+	+	+	+	+	+	+	+	+	+	+	+
Polysaccharides:	Neutral polysaccharides												
	PATAg	+	+	+	+	+	+	−	−	−	+	+	+
	H$_2$O$_2$ + PATAg	−	−	−	−	−	−	+	+	+	+	+	+
	Acid polysaccharides												
	PTA-HCl	+	−	−	−	−	−	+	+	+	+	+	+
	PTA-CrO$_3$	+	−	−	−	−	−	+	+	+	+	+	+
	Pectins												
	Ruthenium red	+++	++	+	−	−	−	+	++	+++	−	−	−
	Esterified pectins												
	Ferric hydroxylamin	+	+	+	+	+	+	+	+	+	−	−	+
Phenols	SP	+	+	+	+	+	+	+	+	+	+	+	+
Size (µm)		0.9	0.85	0.9	0.01	0.01	0.01	0.3	0.65	0.5	0.01	0.01	0.01

Total size of orbicule (µm): stage 1: 1.5   stage 2: 2.2   stage 3: 2

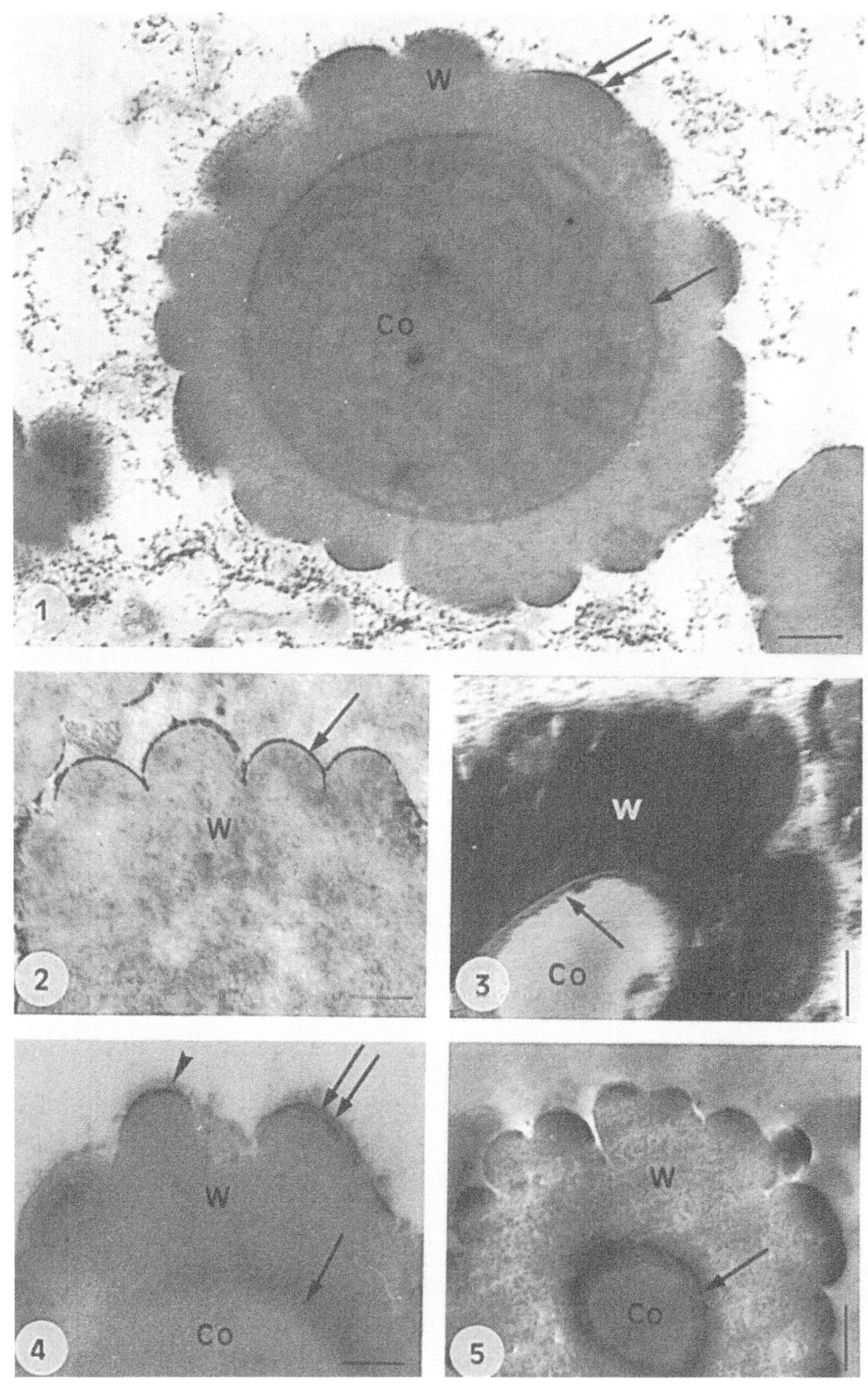

Orbicular core. The central zone of the Ubisch body is not birefringent in polarized light and appears homogeneous in TEM. Proteins are detected by PTA acetone only if the fixative includes OsO_4. These polypeptides are removed by pronase (Fig. 2). Lipids, neutral polysaccharides (Fig. 1), and phenolic compounds are present at each developmental stage. Acid polysaccharides characterized by Ruthenium red, PTA low pH, and PTA chromic acid decrease progressively from stage 1 to stage 3 (Table 1). After acetolysis, none of the cytochemical tests was positive.

Orbicular core-wall interface. It is clearly revealed by $KMnO_4$ protocol (Fig. 3) as an extremely thin zone (a membrane-like sheet) of about 10 nm thickness (Table 1).

In this zone, pronase sensitive proteins (Figs. 2,4), unsaturated lipids (Fig. 5), esterified pectins (Fig. 6), and polyphenols can be detected but no acid polysaccharides. It should be noted that this membrane-like sheet is also sensitive to acetolysis since no cytochemical test is positive after acetolysis.

Orbicular wall. It is homogeneous in TEM but birefringent in polarized light, showing a regular spatial organization. The wall is resistant to acetolysis but is very sensitive to oxidative treatment by $KMnO_4$ (Fig. 3).

Cytochemical reactions concerning the wall give stable results during the orbicule development. Each organic compound researched for is present (Table 1). The level of pectins increases from stage 1 to stage 3.

Orbicular peripheral sheet. It is a 10 nm thick zone that surrounds the whole Ubisch body and can be found in each stage of orbicule development. It contains pronase-resistant proteins (Figs. 2,4). Some of them are organized in filaments perpendicular to the orbicular wall surface (Fig. 4). Acid polysaccharides revealed by PTA low pH and PTA chromic acid (Fig. 7) show the same feature as the filamentous proteins (Fig. 10). Unsaturated lipids (Fig. 8) and neutral polysaccharides (Fig. 9) were also detected. In orbicular tangential sections the peripheral sheet appears as a PATAg positive network of polygons (Fig. 11).

◀ ──

Figs. 1–5. Ubisch bodies in *Lilium*. – Fig. 1. Stage 1. GOBA. PATAg test. Middle section of an orbicule showing the orbicular core (Co), an orbicular core-wall interface (arrow), the orbicular wall (W) and an orbicular peripheral sheet (double arrow). Bar: 0.2 µm. – Fig. 2. Stage 3. GOBA. PTA acetone reaction after pronase treatment. Pronase removes proteins from each orbicular zone except from the peripheral sheet (arrow). Bar: 0.2 µm. Fig. 2 is part of a figure from CLÉMENT & AUDRAN (1993a). – Fig. 3. $KMnO_4$. Without staining. The orbicular core-wall interface is shown by potassium permanganate (arrow) whereas the orbicular wall is partially degraded. Bar: 0.2 µm. Fig. 3 is part of a figure from CLÉMENT & AUDRAN (1993b). – Fig. 4. Stage 3. GOBA. PTA acetone test. Proteins are localized on the core-wall interface (arrow) on the peripheral sheet (double arrow). Moreover, filamentous proteins are detected perpendicular to the orbicular peripheral sheet (arrowhead). Bar: 0.2 µm. – Fig. 5. Stage 1. OsO_4. Without staining. Orbicular core-wall interface is osmiophilic (arrow). Bar: 0.15 µm

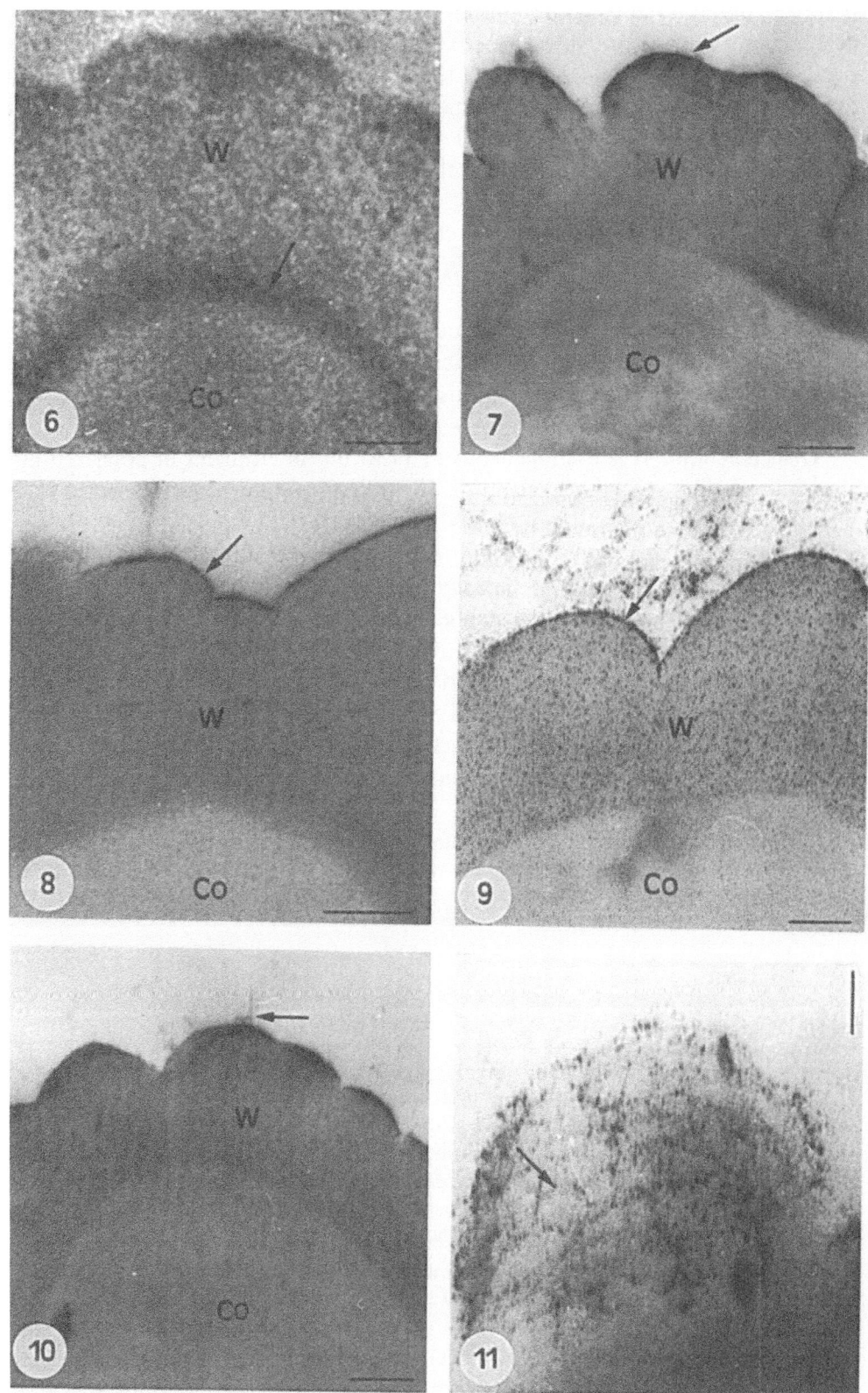

Discussion

Orbicular core. The absence of birefringence within this zone indicates that its molecules are dispersed isotropically. The orbicular core contains unsaturated lipids as derivatives from proorbicules which are composed essentially by tapetal fatty acids (EL-GHAZALY & JENSEN 1987).

Occurrence of proteins and polysaccharides such as pectins is rarely described in the literature. Pectins were found in *Pinus* Ubisch bodies (ROWLEY & WALLES 1987). Proteins seem to be very instable: OsO_4 binds to proteins (ROWLEY & EL-GHAZALY 1992) and seems to be necessary to maintain them through fixation. They remain detectable by PTA acetone and are easily removed by pronase. This suggests that the core proteins exist in a solubilized state with a low concentration. Enzymes involved in sporopollenin biosynthesis around the proorbicule could be synthesized in the tapetal cytoplasm and stored in the proorbicule, they become activated after the proorbicule is extruded in the loculus.

The decrease of acid polysaccharides in the orbicular core of *Lilium* during parallel growth of orbicular wall can be interpreted as follows. Pectins are structural polymers; they concurrently increase in the orbicular wall and so core pectins could be used as skeleton polymers in the construction of the peripheral wall. The orbicular core would play an important role in the orbicular wall ontogenesis. This leads us to an interesting point concerning the orbicular interface.

Orbicular core-wall interface. This area of the Ubisch body was not discussed by previous authors working on *Lilium* (HESLOP-HARRISON & DICKINSON 1969, REZNICKOVA & WILLEMSE 1980). The positive staining by $KMnO_4$ (HAYAT 1970), and the thickness of 10 nm indicate that the orbicular core-wall interface is structurally comparable with a cytomembrane. Cytochemically, the presence of proteins and of unsaturated lipids as far as the absence of neutral polysaccharides is in agreement with this hypothesis. Native pectins are not detected but esterified ones are localized instead. A passage of acid polysaccharides from the core towards the wall would suggest that pectins are initially settled in this membrane-like sheet and then esterified during sporopollenin synthesis. The disappearance of this structure by acetolysis clearly shows that the membrane-like sheet is not formed by sporopollenin.

◄ ───────────────────────────────────────

Fig. 6–11. Ubisch bodies in *Lilium*. – Fig. 6. Stage 1. GFH. Without staining. Esterified pectins revealed by ferric hydroxylamin are visualized on the core-wall interface (arrow). Bar: 0.1 μm. Fig. 6 is part of a figure from CLÉMENT & AUDRAN (1993a). – Fig. 7. Stage 2. GOBA. PTA chromic acid test. Acid polysaccharides are detected on the surface of the orbicular wall (arrow). Bar: 0.3 μm. – Fig. 8. Stage 3. GOBA. TSC-SP reaction. Unsaturated lipids are particularly well contrasted on the orbicular peripheral sheet (arrow). Bar: 0.2 μm. – Fig. 9. Stage 1. GOBA. H_2O_2 + PATAg reaction (control for Fig. 8, avoiding OsO_4). Silver grains are dispersed in the wall and concentrated on the peripheral sheet (arrow). Bar: 0.1 μm. – Fig. 10. Stage 2. GOBA. PTA chromic acid test. This test for acid polysaccharides is positive for filaments oriented perpendicularly to the orbicular wall (arrow). Bar: 0.3 μm. – Fig. 11. Stage 1. GOBA. PATAg reaction. Surface tangential section of an orbicule showing that the peripheral sheet is organized in a polygonal network (arrow). Bar: 10 nm.

The orbicular core-wall interface is a biological membrane surrounding the proorbicule (EL-GHAZALY & NILSSON 1991) and is coated with a glycocalyx on which sporopollenin first polymerises (ROWLEY & SKVARLA 1974, ROWLEY & SRIVASTAVA 1986). It is an ideal place to control substance exchange (proteins, pectins) between the orbicular core and the wall, and also the pathway of sporopollenin biosynthesis.

Orbicular wall. The orbicular wall, unlike the core zone, is birefringent proving that it is structurally well organized (LACEY 1989). The size of the orbicular wall is an important aspect (first increase and then decrease): it was presumed that the orbicule represents a transitory reservoir of sporopollenin for the pollen grain wall (HESLOP-HARRISON 1968, RISUEÑO & al. 1969, ECHLIN 1971, BHANDARI 1984). But exines also show similar variations (REZNICKOVA & WILLEMSE 1980, EL-GHAZALY & JENSEN 1986), so it can be supposed that pollen grain and orbicular walls undergo simultaneously the same events: first sporopollenin is synthesized on both proorbicule and microspore which leads to a thickness increase; then, during maturation, sporopollenin may be compressed inducing a reduction in the size of both walls (CLAUGHER & ROWLEY 1987).

Proteins are easier to preserve in the wall than in the core. They are probably integrated in the sporopollenin construction. The presence of unsaturated lipids revealed by OsO_4 and TSC-SP is in agreement with biochemical analyses. Sporopollenin is supposed to contain an important fraction of lipids (SOUTHWORTH 1990).

Ultrathin sections need to be previously treated by H_2O_2 as a control for the significance of the PATAg reaction. This means that OsO_4, removed by H_2O_2 and fixed on double bonds of the aliphatic carbons, occupies vic-glycol groups, target points of the PATAg test. From this is suggested that lipids and polysaccharides are closely associated in the orbicule sporopollenin as glycolipids (ROWLEY 1985).

Polyphenols represent 19% of purified exines (SCHULZE OSTHOFF & WIERMANN 1987), and several specific phenolic acids have been identified by SOUTHWORTH (1974), HERMINGHAUS & al. (1988), WEHLING & al. (1989). Moreover, the occurrence of PAL (Phenyl Ammonia Lyase) activity has been demonstrated in the pathway of pollen sporopollenin synthesis (RITTSCHER & WIERMANN 1988, BEERHUES & al. 1993). So a positive reaction of the orbicular wall with polyphenols test is not surprising and is a supplementary common feature with pollen walls.

Acetolysis does not affect the apparent structure of orbicular sporopollenin features but oxidative treatment with $KMnO_4$ induces degradation of the orbicular wall. These two features were also found in pollen sporopollenin (ROWLEY & PRIJANTO 1977, AUDRAN 1978).

In our studies each feature of the orbicular wall is identical to that of the pollen grain exine. ROWLEY & al. (1959) pointed out that the Ubisch body ornamentation is identical with that of the exine in *Poa* (*Poaceae*), *Degeneria* (*Degeneriaceae*), and *Cryptomeria* (*Pinaceae*) pollen. In accordance with EL-GHAZALY & JENSEN (1987) we conclude that the orbicular sporopollenin is elaborated at the same time and in the same way than the pollen one.

Orbicular peripheral sheet. Substructural organization of pollen sporopollenin in *Lilium* (SOUTHWORTH 1985) and in *Centrolepis aristata* (ROWLEY & DUNBAR 1990) is described as a polygonal framework. In *Lilium*, the orbicules also show polygonal features on their surface. The polygonal appearance of the orbicular surface may be explained according to the model of ROWLEY (1990). Orbicular sporopollenin then should be of the same architecture as that of the pollen grain exine.

Proteins are not affected by pronase treatment suggesting that they are strongly settled in the orbicular wall. Filamentous proteins and acid polysaccharides together form a glycocalyx structure on the surface of the Ubisch body. These filaments represent remnants of the glycocalyx supported by the membrane that surrounds the Ubisch body's core (ROWLEY & SKVARLA 1974) and on which sporopollenin polymerisation occurs first (WATERKEYN & BIENFAIT 1971). Glycoprotein filaments elongate during development of the orbicular wall.

Each organic compound revealed in the orbicular wall is also detected on the peripheral sheet except for pectins. A particular peripheral structure was already studied on pollen grain exine in *Brassica* (GAUDE & DUMAS 1984) and in *Apium* (WEBER 1992) and was suggested to be a biological membrane. Our results do not allow us to conclude that the orbicular peripheral sheet is a membranous structure but only a special zone with sporopollenin biosynthesis.

Conclusions

Ubisch bodies (orbicules) appear to be composed of four concentric interdependant structures (Fig. 12). The orbicular core is a sphere containing a complex mixture of organic compounds surrounded by a membrane-like sheet. On this membrane a glycocalyx is erected and sporopollenin is elaborated on its glycoprotein filaments. The glycocalyx grows progressively until the entire orbicular wall is finished. The wall includes two kinds of compounds: 1. structural elements (proteins, pectins, polysaccharides and glycolipids) organized in a tridimensional polygonal frame (ROWLEY 1990), and 2. filling elements (unsaturated lipids and polyphenolic compounds). The former can be seen on the surface of the Ubisch body. Effectively, on this site, architectural molecules are already deposited but not yet embedded with lipids and polyphenols and can then be visualized with cytochemical tests.

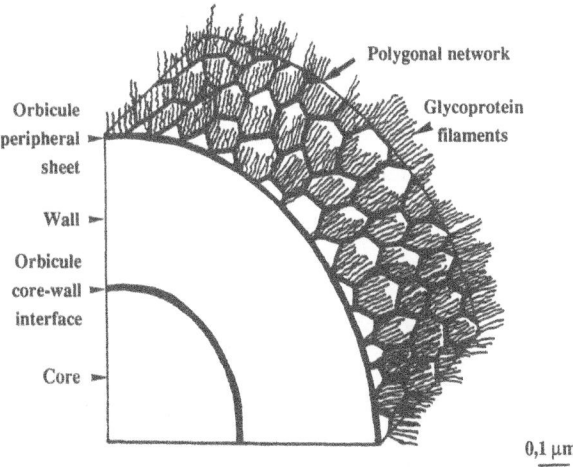

Fig. 12. Model of orbicular sporopollenin architecture and biosynthesis

The control of exine biosynthesis through the plasmalemma is modulated continuously by the cytoplasm of the microspore. It is very different in the case of Ubisch body. Effectively, the membrane that surrounds the orbicular core is synthesized within the tapetal cytoplasm and any modifications may occur in this membrane since the proorbicule is secreted towards the locular fluid.

We thank Dr JOHN ROWLEY for helpful comments on the manuscript.

References

AUDRAN, J. C., 1978: Présentation de quelques transformations structurales et texturales des exines de *Cycadales* traitées par l'acétolyse puis fixées par le permanganate de potassium. – Ann. Mines Belgique **2**: 133–141.

BANERJEE, U. C., BARGHOORN, E. S., 1971: The tapetal membranes in grasses and Ubisch body control of mature exine pattern. – In HESLOP-HARRISON, J., (Ed.): Pollen: development and physiology, pp. 126–127. – London: Academic Press.

BENEDETTI, E. L., BERTOLINI, B., 1963: The use of phosphotungstic acid as a stain for the plasma membrane. – J. Roy. Microsc. Soc. London **81**: 219–222.

BEERHUES, L., RITTSCHER, M., SCHÖPKER, H., SCHWERDTFEGER, C., WIERMANN, R., 1993: The significance of the tapetum in the biochemistry of pollen pigmentation – an overview. Pl. Syst. Evol. [Suppl.] **7**: 117–125.

BHANDARI, N. N., 1984: The microsporangium. – In JOHRI, B. M., (Ed.): Embryology of angiosperms, pp. 53–122. – Berlin: Springer

CLAUGHER, D., ROWLEY, J. R., 1987: *Betula* pollen grain substructure revealed by fast atom etching. – Pollen & Spores **29**: 5–20.

CLÉMENT, C., AUDRAN, J. C., 1992: Apports de la cytochimie à la connaissance des orbicules dans l'anthère de *Lilium* (Liliacées). 1 – Le coeur orbiculaire. – Bull. Soc. Bot. Fr., **189**, Lettres bot. (4/5), 369–376.

–   – 1993 a: Electron microscopical evidence of a membrane around Ubisch body's core in *Lilium* (*Liliaceae*). – Grana **32**: 311–314.

–   – 1993 b: Orbicule wall surface characteristics in *Lilium* (*Liliaceae*). An ultrastructural and cytochemical approach. – Grana (in press).

ECHLIN, P., 1971: The role of the tapetum during microsporogenesis of angiosperms. – In HESLOP-HARRISON, J., (Ed.): Pollen: development and physiology, pp. 41–61. – London: Butterworths.

EL-GHAZALY, G., JENSEN, W. A., 1986: Studies of wheat (*Triticum aestivum*) pollen. 1. Formation of the pollen wall and Ubisch bodies. – Grana **25**: 1–25.

–   – 1987: Development of wheat (*Triticum aestivum*) pollen. 2. Histochemical differentiation of wall and Ubisch bodies during development. – Amer. J. Bot. **74**: 1396–1418.

–   NILSSON, S., 1991: Development of tapetum and orbicules of *Catharanthus roseus* (*Apocynaceae*). – In BLACKMORE, S., BARNES, S. H., (Eds): Pollen and spores, pp. 317–325. – Oxford: Clarendon Press.

ERDTMAN, G., 1960: The acetolysis method. – Svensk Bot. Tidskr. **54**: 561–564.

GAUDE, T., DUMAS, C., 1984: A membrane-like structure on the pollen wall surface in *Brassica*. – Ann. Bot. **54**: 821–825.

HAYAT, M. A., 1970: Principles and techniques of electron microscopy. 1: Biological applications. – New York: Van Nostrand Reinhold.

HERMINGHAUS, S., ARENDT, S., GUBATZ, S., RITTSCHER, M., WIERMANN, R., 1988: Aspects of sporopollenin biosynthesis: phenols as integrated compounds of the biopolymer. – In CRESTI, M., GORI, P., PACINI, E., (Eds): Sexual reproduction in higher plants, pp. 169–174. – Berlin: Springer.

HESLOP-HARRISON, J., 1968: Pollen wall development. – Science **161**: 230–237.

– DICKINSON, H. G., 1969: Time relationships of sporopollenin synthesis associated with tapetum and microspores in *Lilium*. – Planta **84**: 199–214.

KEIJZER, C. J., 1986: The processes of anther dehiscence and pollen dispersal. 2. The formation and the transfer mechanism of pollenkitt, cell wall development in the locule tissues and a possible function of the orbicules in pollen dispersal pp. 31–47. – Thesis, University of Wageningen.

MARINOZZI, V., LUZZATTO, A. C., DEVIRGILIIS, L. C., MAZZATENTA, C., 1977: Osmium dependent argentaffinity of elastic fibers: its relations with polyphenolic compounds. – J. Submicrosc. Cytol. **9**: 267–274.

LACEY, A. J., 1989: Light microscopy in biology. A practical approach. – Oxford, New York, Tokyo: Irl Press.

PACINI, E., FRANCHI, G. G., 1991: Diversification and evolution of the tapetum. – In BLACKMORE, S., BARNES, S. H., (Eds): Pollen et spores, pp. 301–316. – Oxford: Clarendon Press.

RAJ, B., EL-GHAZALY, G. A., 1987: Morphology and taxonomic application of orbicules (Ubisch bodies) in *Chloanthaceae*. – Pollen et Spores **29**: 151–166.

RAMBOURG, A., 1969: Localisation ultrastructurale et nature du matériel coloré au niveau de la surface cellulaire par le mélange chromique phosphotungstique. – J. Microscopie **8**: 325–342.

REZNICKOVA, S. A., WILLEMSE, M. T. M., 1980: Formation of pollen in the anther of *Lilium*. 2. The function of the surrounding tissues in the formation of pollen and pollen wall. – Acta Bot. Neerl. **29**: 141–156.

– VAN AELST, A. C., WILLEMSE, M. T. M., 1980: Investigation of exine and orbicule formation in the *Lilium* anther by scanning electron microscopy. – Acta Bot. Neerl. **29**: 157–164.

RISUEÑO, M. C., GIMENEZ, G., LOPEZ-SAEZ, J. F., GARCIA, M. I. R., 1969: Origin and development of sporopollenin bodies. – Protoplasma **67**: 361–374.

RITTSCHER, M., WIERMANN, R., 1988: Studies on sporopollenin biosynthesis in *Tulipa* anthers. 2. Incorporation of precursors and degradation of the radiolabelled polymers. – Sex. Pl. Reprod. **1**: 132–139.

ROLAND, J. C., 1978: General preparation and staining of thin sections. – In HALL, J. L., (Ed.): Electron microscopy and cytochemistry of plant cells, pp. 1–62. – Amsterdam, Oxford, New York: Elsevier/North-Holland Biomedical Press.

ROWLEY, J. R., 1975: Lipopolysaccharides embedded within the exine of pollen grains. – In BAILEY, G. W., (Ed.): 33rd Proc. Electron Microscopy Soc. Amer., pp. 572–573.

– 1990: The fundamental structure of pollen exine. – Pl. Syst. Evol. [Suppl.] **5**: 13–29.

– DUNBAR, L., 1990: Outward extension of spinules in exine of *Centrolepis aristata* (*Centrolepidaceae*). – Bot. Acta **103**: 355–359.

– EL-GHAZALY, G., 1992: Lipid in wall and cytoplasm of *Solidago* pollen. – Grana **31**: 273–283.

– MÜHLETHALER, K., FREY-WYSSLING, A., 1959: A route for the transfer of materials through the pollen grain wall. – J. Biophys. Biochem. Cytol. **6**: 537–538.

– PRIJANTO, B., 1977: Selective destruction of the exine of pollen grains. – Geophytology **7**: 1–23.

– SKVARLA, J. J., 1974: Plasma membrane-glycocalyx origin of Ubisch body wall. – Pollen & Spores **16**: 441–448.

– SRIVASTAVA, K. K., 1986: Fine structure of *Classopollis* exines. – Canad. J. Bot. **64**: 3059–3074.

– WALLES, B., 1987: Origin and structure of Ubisch bodies in *Pinus sylvestris*. – Acta Soc. Bot. Polon. **56**: 215–227.

Schulze Osthoff, K., Wiermann, R., 1987: Phenols as integrated compounds of sporopollenin from *Pinus* pollen. – J. Pl. Physiol. **131**: 5–15.

Seligman, A. M., Wasserkrug, H. L., Hanker, J. S., 1966: A new staining method (OTO) for enhancing contrast of lipid containing membranes and droplets in osmium-tetroxide fixed tissue with osmiophilic thiocarbo-hydrazide (TCH). – J. Cell Biol. **30**: 424–432.

Southworth, D., 1974: Solubility of pollen exines. – Amer. J. Bot. **61**: 36–44.

– 1985: Pollen exine substructure. 1. *Lilium longiflorum*. – Amer. J. Bot. **72**: 1274–1283.

– 1990: Exine biochemistry. – In Blackmore, S., Knox, R. B., (Eds): Microspores: evolution and ontogeny, pp. 193–212. – London: Academic Press.

Thiéry, J. P., 1967: Mise en évidence des polysaccharides sur coupes fines en microscopie electronique. – J. Microscopie **6**: 987–1018.

– Rambourg, A., 1974: Cytochimie des polysaccharides. – J. Micoscopie **21**: 225–232.

Waterkeyn, L., Bienfait, A., 1971: Primuline-induced fluorescence of the first exine elements and Ubisch bodies in *Ipomoea* and *Lilium*. – In Brooks, J., Grant, P. R., Muir, M. D., Van Gijzel, P., Shaw, G., (Eds): Sporopollenin, pp. 108–129. – London, New York: Academic Press.

Weber, M., 1992: Nature and distribution of the exine-held material in mature pollen grains of *Apium nodiflorum* L. (*Apiaceae*). – Grana **31**: 17–24.

Wehling, K., Niester, C., Boon, J. J., Willemse, M. T. M., Wiermann, R., 1989: p-coumaric acid: a monomer in the sporopollenin skeleton. – Planta **179**: 376–380.

Address of authors: Christophe Clément, Jean-Claude Audran, Laboratoire de Biologie et Physiologie Végétales de l'UR4R, P.B. 347, F-51062 Reims Cedex, France.

Pl. Syst. Evol. [Suppl.] 7: 75–90 (1993)

Immunolocalization of nuclear antigens and ultrastructural cytochemistry on tapetal cells of *Scilla peruviana* and *Capsicum annuum*

Pilar S. Testillano, Pablo Gonzalez-Melendi, Begoña Fadon, Amelia Sanchez-Pina, Adela Olmedilla, and Maria del Carmen Risueño

Key words: *Hyacinthaceae, Scilla peruviana, Solanaceae, Capsicum annuum.* – Secretory tapetum, immunofluorescence, immunogold labelling, nucleus, nucleolus, chromatin, interchromatin region, nuclear antigens, NAMA-Ur method.

Abstract: The nuclei of the secretory tapetum of *Scilla peruviana* L. and *Capsicum annuum* L. are studied by light and transmission electron microscopy using cryoprocessed material. The NAMA-Ur method, specific for DNA, anti-DNA and anti-histones immunogold labelling show the condensed chromatin pattern and reveal the presence of chromatin fibres in the interchromatin region. Different antigens of the nucleoplasmic snRNPs, involved in the splicing of the pre-mRNAs, are immunolocalized over the interchromatin fibres. Nucleolar proteins related to different steps of rRNA processing are immunodetected on the nucleolar components. These approaches give new insights in the nuclear function of the tapetum during pollen development.

The tapetum is the innermost layer of the anther wall that surrounds the sporogenous tissue. Tapetal cells are of crucial importance in nourishing the spore mother cells and in contributing to the formation of the exine and the deposition of the tryphine and pollenkitt (Bhandari 1984, Shivanna & Johri 1985). This key role is illustrated by the fact that degeneration of pollen grains in many male sterile lines is invariably associated with the malfunctioning of the tapetum (Horner & Rogers 1974, Shivanna & Johri 1985, Albertini & al. 1987, Majewska-Sawka & al. 1990).

Two types of tapetum, secretory and plasmodial are known, but fine structural studies have mainly been conducted on the secretory type (Bhandari 1984, Shivanna & Johri 1985). Generally, the secretory tapetum is formed by a single layer of large cells which undergo changes in activity related to their function during microsporogenesis. This dynamism involves specific structural features and ends with the degeneration of the tapetum before anthesis. Even though the tapetum has been investigated in many aspects (reviewed in Chapman 1987), the mostly lightmicroscopic data available about the tapetal nuclei originate from the sixties or the seventies (Carniel 1963, Buss & Lersten 1975), only few papers with ultrastructural data have been published in the last decade (e.g. Risueño & Medina 1986, Albertini & al. 1987).

In the last few years many investigations have pursued the localization of the nuclear structures of different molecules playing important roles in the nuclear function, as well as the cytochemical and ultrastructural characterization of these structures. These studies have revealed the presence of DNA, different RNAs and different proteins in specific nuclear domains; this illustrates the widely accepted feature of the functional compartmentalization of the nucleus (Harris & Zbarsky 1990, Jackson 1991).

The functional compartmentalization of the nucleus correlates with the state of nuclear activity that finally regulates the whole cellular function (Jackson 1991). The condensation state of the chromatin, the presence of different structures in the interchromatin region and the nucleolar organization show dramatic changes during plant developmental processes (Medina & al. 1983a, Risueño & Medina 1986, Risueño & al. 1988, Testillano & Risueño 1988, Risueño 1990).

Low temperature methods of processing preserve the cellular structures closer to the in vivo state than conventional ones, and enable the use of different cytochemical methods (Steinbrecht & Zierold 1987). These cryomethods have so far been only rarely applied to plant cells and, especially, to anthers (Sanchez-Pina & al. 1989, 1990; Alche & al. 1990; Testillano & al. 1991a, b, 1992a; Fitzgerald & al. 1993, Hesse & Hess 1993).

In the present work the nuclei of tapetal cells are studied using cryoprocessed material and different cytochemical and immunocytochemical methods at light and electron microscopic level.

Material and methods

Material. Plants of *Capsicum annuum* L. (*Solanaceae*) and *Scilla peruviana* L. (*Hyacinthaceae*) were grown in a greenhouse with controlled temperature and photoperiod. Flower buds bearing anthers at the developmental stages ranging from meiosis to first pollen mitosis were selected for investigation.

Conventional processing for light and electron microscopy. Anthers were carefully excised from flower buds and fixed in 3% glutaraldehyde in 0.025 M cacodylate buffer for 4 h. After washing in the same buffer some of them were postfixed in 1% Osmium tetroxide and washed again. Finally, the samples were dehydrated in an ethanol series and embedded in Epon. Semithin sections were observed under phase contrast without any staining for general structure observation. Ultrathin sections were mounted on pioloform-coated copper grids and stained with uranyl acetate and lead citrate.

Special processing methods. 1. Curcumine staining (Stockert & al. 1989). Semithin sections from glutaraldehyde fixed and Epon embedded samples were stained with 0.2 mg/ml curcumine in 5% ethanol for 30 min. Viewed under violet-blue irradiation ($\lambda = 436$ nm) in a Zeiss fluorescence photomicroscope the nuclei and the cell walls could be observed.

2. Lowicryl embedding. Anthers were fixed in 4% formaldehyde in PBS for 18 h at 4 °C. After washing in PBS, they were dehydrated in a methanol series at 4 °C and embedded in Lowicryl K4M at 4 °C under UV irradiation. Lowicryl ultrathin sections were mounted on pioloform and carbon-coated either on copper or gold grids and used for cytochemistry and immunogold labelling.

3. Cryofixation and cryoultramicrotomy. Anthers were prefixed in formaldehyde, cryoprotected and cryofixed in liquid propane at −190 °C, as previously described

(TESTILLANO & al. 1992a). 1 μm cryosections were placed on slides, covered by sucrose and stored at −20 °C until use for immunofluorescence.

Antibodies. The antibodies used and their dilutions were: Anti-DNA, mouse monoclonal IgM against double and single stranded DNA (Boehringer Mannheim); dilution used: 25 μg/ml. Anti-histone H2B, rabbit serum (MÜLLER & al. 1991); dilution 1/600. Anti-snRNPs: two mouse monoclonals IgG recognizing common proteins of the nucleoplasmic snRNPs, the 2X.10H3.18.10 and 1X.5C10.1.2 (LUTZ 1986); they were used undiluted. Anti-m3G, rabbit polyclonal IgG against the tri-methyl-guanosine cap of all snRNPs (LÜHRMANN & al. 1982), dilution 10 mg/ml. Anti-B36, mouse monoclonal IgG against the B36 nucleolar protein (CHRISTENSEN & al. 1986) dilution 1/5. D77, mouse monoclonal IgG against the nucleolar protein fibrillarin (ARIS & BLOBEL 1988), it was used undiluted. J26, human sera against 50 and 86 KD nucleolar proteins (VERHEIJEN & al. 1986), it was used undiluted.

Immunogold labelling. Lowicryl ultrathin sections were treated as previously described (TESTILLANO & al. 1992a) for all antibodies except for the anti-DNA. Shortly the grids were floated on distilled water, PBS, 5% BSA, the first antibody, PBS, the second antibody conjugated to 10 or 15 nm gold particles, PBS and water. In the case of anti-DNA, a 1/50 (v/v) solution of goat normal serum in PBS was used instead of PBS for the washes and the antibody dilutions. Finally, the sections were stained with one of the following staining methods: uranyl and lead, EDTA or the NAMA-Ur method. Controls were carried out by omitting the first antibody.

Ultrastructural cytochemistry. For RNPs: the EDTA procedure (BERNHARD 1969) was followed on Lowicryl ultrathin sections. For DNA: the NAMA-Ur method (TESTILLANO & al. 1991b) was performed on gold grids carrying Lowicryl ultrathin sections after the anti-DNA immunodetection.

Results

The secretory tapetum of *Scilla peruviana* and *Capsicum annuum* forms one peripheral layer of large cells in the locule (Figs. 1 and 2). These large tapetal cells undergo mitosis without cytokinesis during pachytene of meiocytes and appear binucleate there after and throughout the postmeiotic interphase showing two large round nuclei (Figs. 1 and 2). The tapetal tissue persists through microsporogenesis (Fig. 2) until the late vacuolate microspore stage, when it degenerates and tapetal cells only show dense lipid bodies and some cytoplasmic remnants (Fig. 3). In this period the cytoplasm of the tapetal cells shows many ribosomes and abundant rough endoplasmic reticulum. Also, several large vacuoles can be seen in each tapetal cell (Figs. 2 and 4).

Ultrastructure of the tapetal nuclei. The tapetal nuclei have specific features, different from the sporogenous tissue and the other tissues forming the anther. At the ultrastructural level, these nuclei show a lobed outline with several digitations (Fig. 4). The nucleus consists of condensed chromatin masses and shows less dense fibrillo-granular structures (Fig. 4).

Each nucleus has one or two small spherical nucleoli. These have two ultrastructurally distinct appearances: the segregated one that is typical of low activity stages, and the compact one exclusively formed by a dense fibrillar component (F). The segregated nucleoli show a granular component (G) which surrounds the F (Fig. 4); this nucleolar type appears in tapetal nuclei shortly before the nuclear division. The compact nucleoli are seen in subsequent stages of

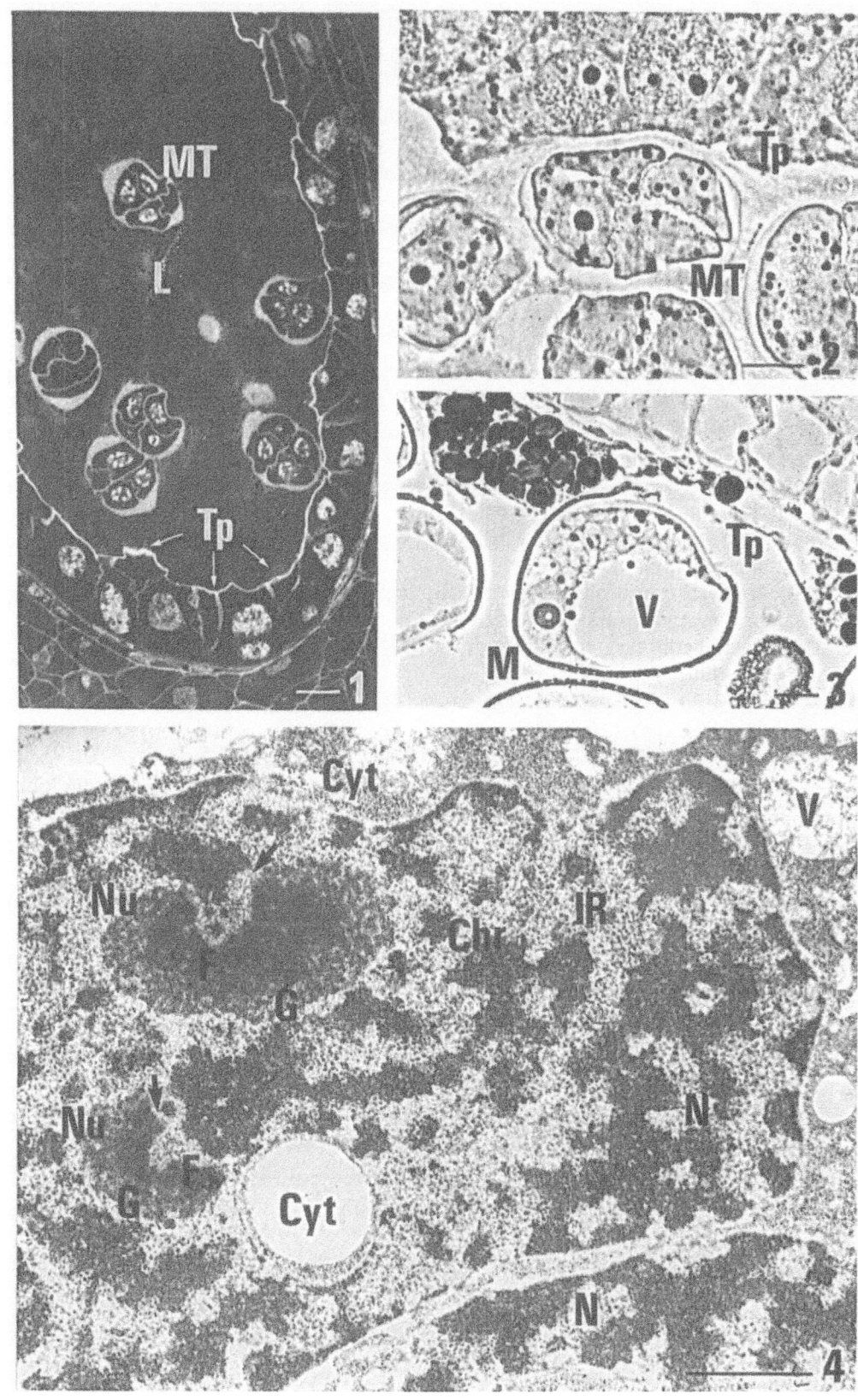

development (Figs. 5, 6, 7, 16 and 17). In all cases, the nucleoli show a unique heterogeneous and large fibrillar centre with condensed chromatin cores (Figs. 4 and 5). The heterogeneous fibrillar centre is connected to the extranucleolar chromatin corresponding to the chromosome bearing the nucleolar organizer region (NOR). This sometimes gives the nucleolus a typical "horse-shoe" morphology observed in many tapetal nuclei (Figs. 4, 5, 14, 16 and 17).

Cytochemistry and immunocytochemistry. The cytochemical and immunocyto-chemical tests provide similar results in tapetal nuclei during all stages of microspore development. The EDTA cytochemical staining, preferential for ribonucleoproteins (RNPs), provides a good contrast for the nucleolus and the interchromatin structures while the condensed chromatin masses appear bleached (Fig. 5). This cytochemical method reveals abundant single or clustered 30–50 nm granules within the IR (Fig. 6). These granules are also evident in conventional EM preparations (Fig. 4). EDTA-positive fibrillar structures of different thickness are also seen forming a network through the entire interchromatin region (Figs. 6 and 16). A densely stained coat of perichromatin fibres is absent at the periphery of the bleached condensed chromatin masses (Figs. 5, 6 and 16). Some RNP nuclear bodies appear in the IR after EDTA staining (Figs. 5 and 6).

Immunolabelling with anti-DNA antibodies provides a dense labelling of the condensed chromatin masses (Figs. 7–9). Several gold particles are also seen attached to fibrillar structures of the IR (Figs. 7 and 9). When the anti-DNA immunodetection is combined with the NAMA-Ur specific staining for DNA, the specifically stained condensed chromatin patches and fibres of dispersed chromatin in the IR are selectively labelled (Figs. 8 and 9). Some gold particles are also localized on the nucleolus (Fig. 8).

With anti-histone H2B antibodies a similar pattern of labelling is obtained for the condensed chromatin masses and some fibres of the IR (Fig. 10).

The antibodies against proteins of the nucleoplasmic small nuclear ribonucleo-protein particles (snRNPs) give positive immunofluorescence to the nuclei in all anther tissues (Fig. 11). The immunofluorescence is not homogeneously distributed throughout the nucleoplasm, but forms a bright fine reticulum with dark regions, and the nucleoli appear negative (Fig. 12). At the EM level, using both antibodies

Figs. 1–4. *Scilla peruviana.* – Fig. 1. Semithin section of anther at tetrad stage. Glutaraldehyde fixation, Epon embedding and Curcumine staining. Under violet-blue irradiation, Curcumine provides fluorescence of the nuclei and the cell walls. *MT* Microspore tetrad, *Tp* Tapetum, and *L* Loculus. Bar: 20 μm. – Figs. 2, 3. Semithin sections of anthers under phase contrast. Glutaraldehyde fixation, Epon embedding. The binucleated tapetal cells appear intact at tetrad (MT) stage (Fig. 2). The tapetal cells (*Tp*) degenerate at the late vacuolate microspore (*M*) stage and many dense bodies are evident (Fig. 3). *V* Vacuole. Bar: 10 μm. – Fig. 4. Tapetal cell, young microspore stage. Glutaraldehyde fixation, Epon embedding, uranyl-lead staining. The two lobed nuclei (*N*) show large condensed chromatin masses (*Chr*) and an interchromatin region (*IR*) rich in granules and fibres; cytoplasmic digitations can be also seen inside it. The two nucleoli (*Nu*) show a segregated morphology with the granular component (*G*) surrounding the dense fibrillar one (*F*) and a large heterogeneous fibrillar centre (arrows). Both nucleoli show a "horse shoe" morphology, i.e. bear the NOR. The cytoplasm (*Cyt*) is rather dense with many ribosomes and vacuoles (*V*). Bar: 1 μm

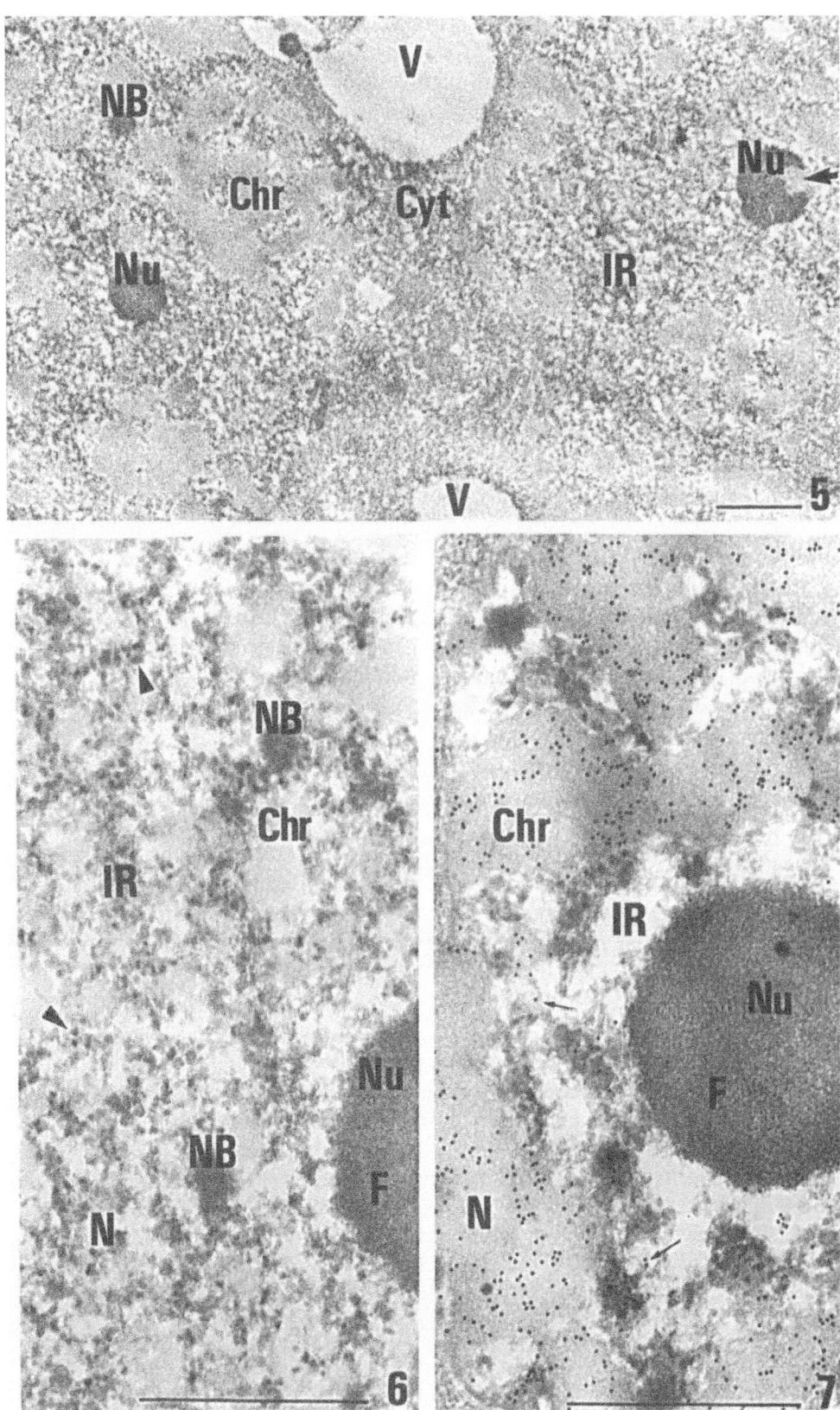

against either proteins of the snRNPs or the m3G-cap of all snRNAs, the nucleoplasmic snRNPs are localized at the interchromatin fibrillar structures which are stained by EDTA, whereas the granules do not show labelling (Figs. 13 and 17). Nuclear bodies are not labelled with anti-snRNPs or anti-m3G antibodies.

Concerning the nucleolar antigens, the J26 antibody directed to two different ribosomal proteins gives positive immunofluorescence to the nucleoli and the cytoplasm of the tapetal cells while the nucleoplasm is negative (Fig. 14). Both antibodies recognizing the two homologous nucleolar proteins fibrillarin and B36 provide similar results. The nucleoli are the only cellular structures positive in the immunofluorescence test (Fig. 15); this fact is consistent with positive immunogold-labelling of the fibrillar component of the nucleolus; the heterogeneous fibrillar centre does not show labelling (Fig. 16). When the anti-m3G antibody against the cap of all snRNAs is used, it also localizes the snRNAs of the nucleolus (snoRNAs) and provides gold labelling of the fibrillar component, the fibrillar centre does not show gold particles (Fig. 17). Controls omitting the first antibody gave no remarkable labelling.

Our results indicate that the tapetal nuclei show a specific chromatin pattern, interchromatin structures, and a nucleolar morphology that are in relation to their state of activity. Factors involved in the splicing of the pre-messenger RNAs and the processing of ribosomal RNAs, as well as DNA and histones are immunolocalized in the tapetal nuclei.

Discussion

The secretory tapetum of *Scilla peruviana* and *Capsicum annuum* shows general morphological features described for other species during pollen development (BHANDARI 1984, SHIVANNA & JOHRI 1985, KEIJZER 1987, CHAPMAN 1987). Tapetum development involves a dedifferentiation until tapetum nuclear mitosis and a subsequent redifferentiation to form a secretory tissue during microsporogenesis; this ends with the degeneration of the tapetal cells at late stages of pollen development, when only large lipid droplets and some cytoplasmic remnants are left. The presence of an abundant endoplasmic reticulum, ribosomes, and small vacuoles is related to the secretory function of the tapetal cells described in other species (CHAPMAN 1987). These activities are regulated by the nuclei of the tapetal cells. The chromatin pattern showed by the tapetal nuclei of *Scilla* and *Capsicum* is

◀ ──

Figs. 5–7. *Capsicum annuum*. Tapetal cells in early vacuolated stage. – Fig. 5. Formaldehyde fixation, Lowicryl embedding and EDTA staining. The ribonucleoprotein structures of the interchromatin region (*IR*) are contrasted as well as the nucleoli (*Nu*) and the nuclear bodies (*NB*); chromatin patches (*Chr*) appear bleached, without densely stained coat of perichromatin fibers (see also Fig. 13). – Fig. 6. At high magnification EDTA stained granules (arrowheads) and fibrillar structures are clearly seen. *N* nucleus, *F* nucleolar dense fibrillar component, *Cyt* cytoplasm and *V* vacuole. – Fig. 7. Nuclear region. Formaldehyde fixation, Lowicryl embedding and anti-DNA immunogold labelling with 15 nm gold particles. Labelling appears on the chromatin (*Chr*) and some fibers (arrows) of the interchromatin region (*IR*). *N* Nucleus, *Nu* nucleolus, and *F* dense fibrillar component. – Bars: 1 μm

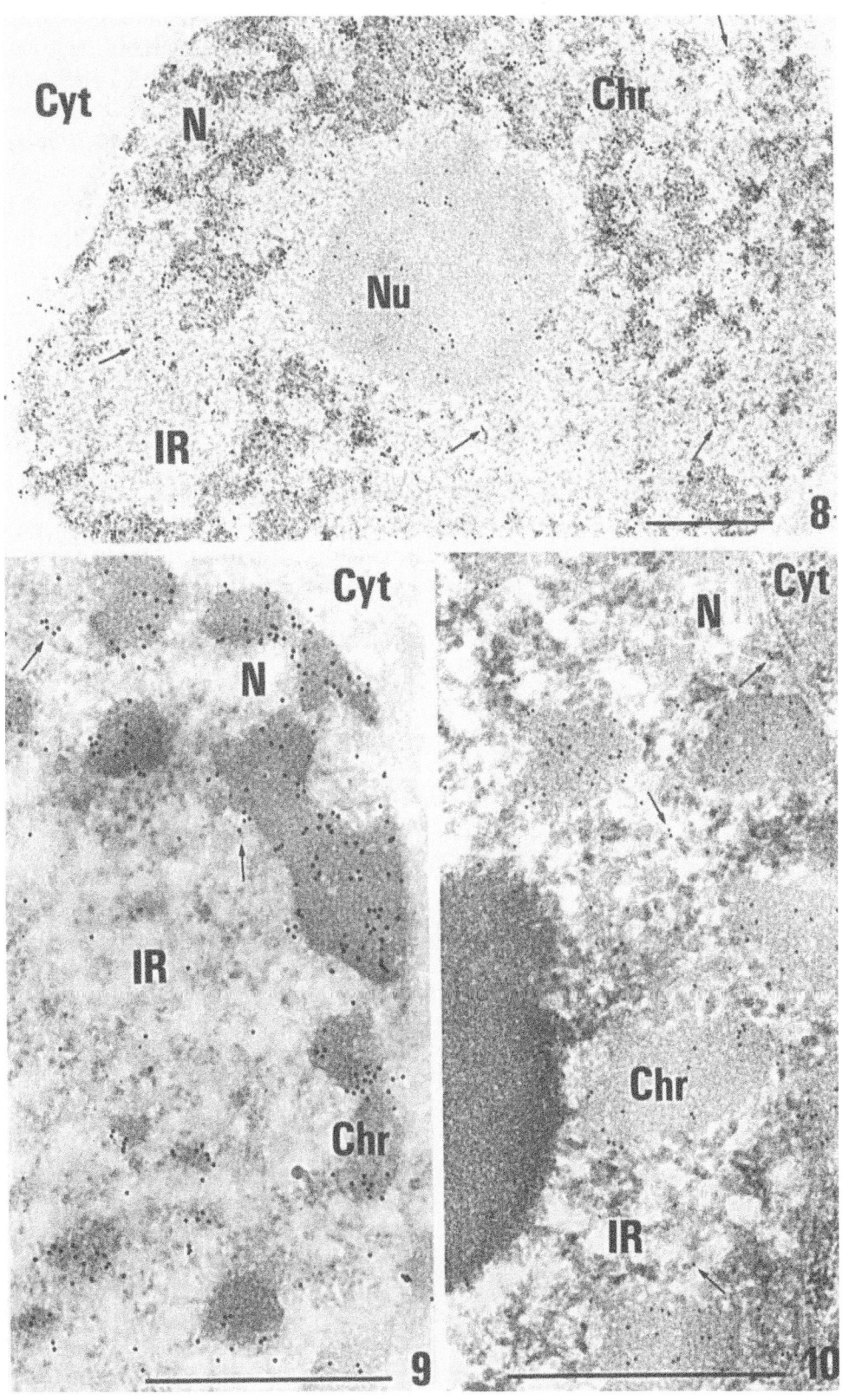

in relation to its state of activity. At early stages of pollen development the condensed chromatin masses show many fibres at their periphery which is a feature of the meristematic nuclei (RISUEÑO & MORENO 1979), and can be related to the dedifferentiation process of the tapetum at these stages (CHAPMAN 1987).

The application of low-temperature methods of processing (TESTILLANO & al. 1991b, 1992a, b) permits the execution of various cytochemical and immunocyto-chemical techniques in the anthers. The EDTA cytochemical method (BERNHARD 1969) preferentially stains the ribonucleoprotein (RNP) structures while condensed chromatin masses are bleached. This staining method allows the observation of the different structures that are composed by ribonucleoproteins; among them, the perichromatin fibres are clearly seen as a dark coating of the condensed chromatin patches. These perichromatin fibres are the ultrastructural counterpart of the newly synthesized RNA and are abundant in transcriptionally highly active nuclei (RISUEÑO & MORENO 1979, FAKAN & PUVION 1980, TESTILLANO & RISUEÑO 1988, RISUEÑO 1990). In subsequent developmental stages the tapetal nuclei show condensed spherical chromatin masses in which not many EDTA-positive peri-chromatin fibres can be detected; this feature is typical for less transcriptionally active nuclei in highly differentiated or quiescent plant tissues (RISUEÑO & MORENO 1979). Other RNP fibres appear in the interchromatin region, they are commonly called interchromatin fibres (FAKAN & PUVION 1980).

.The NAMA-Ur method, specific for DNA (TESTILLANO & al. 1991b), clearly reveals the chromatin patterns of the tapetal nuclei but also stains DNA fibres, corresponding to dispersed chromatin, in the interchromatin region. This fact has not been previously described except for some data on other plant cells such as pollen grains (TESTILLANO & al. 1991b, 1993). The use of anti-DNA and anti-histones immunogold labelling also indicates the presence of chromatin fibres, stained by NAMA-Ur, in the interchromatin region of these nuclei. Positive immuno-labelling illustrates the conservation of such epitopes in the tapetal cells, where no previous nuclear immunolocalizations have been done so far; and it shows the reliability of the processing method in the maintenance of antigenicity and accessibility for the antibodies in this tissue.

Therefore, the fibres observed at the interchromatin region in conventionally stained sections are of different nature. Firstly, there are RNP fibres, revealed by EDTA, that constitute different steps of transcription and/or processing of the extranucleolar RNAs. Secondly, chromatin fibres are also present in the inter-chromatin region as revealed by NAMA-Ur staining and the anti-DNA and anti-histones immunogold labelling.

◄ ───

Figs. 8–10. *Capsicum annuum.* Tapetal cells in middle vacuolated stage. – Fig. 8. Formaldehyde fixation, Lowicryl embedding, NAMA-Ur staining and anti-DNA immunogold labelling. The NAMA-Ur method stains the condensed chromatin masses (*Chr*) which are labelled with 15 nm gold particles. Labelling is also localized over some stained fibres (arrows) of the interchromatin region (*IR*) and over the nucleolus (*Nu*). – Fig. 9. A detail of the nucleus from Fig. 8 at higher magnification. *Cyt* Cytoplasm. – Fig. 10. Nuclear region. Formaldehyde fixation, Lowicryl embedding, and anti-H2B immunogold labelling; 10 nm gold particles. The condensed chromatin masses (*Chr*) and some fibers (arrows) of the interchromatin region (*IR*) are the only structures labelled. *N* Nucleus and *Cyt* cytoplasm. – Bars: 1 μm

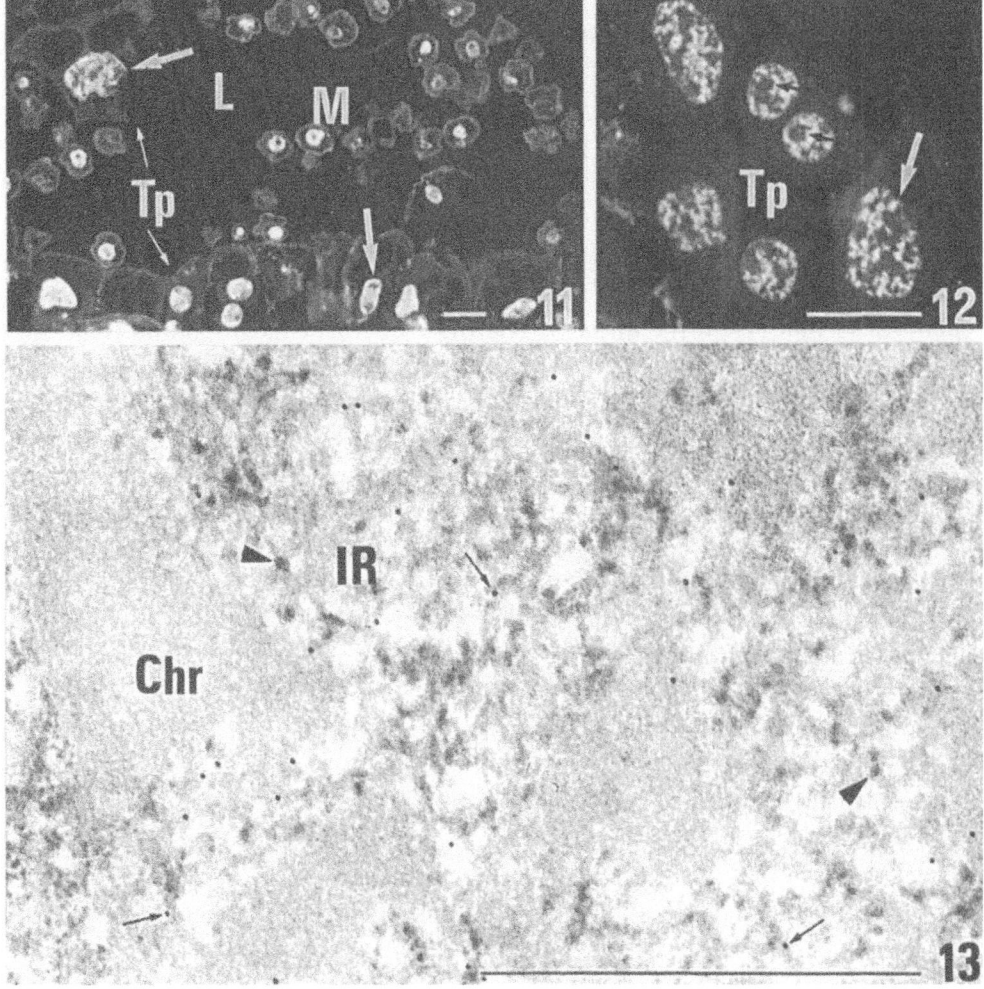

Fig. 11–13. *Capsicum annuum*. Middle microspore stage. – Figs. 11, 12. Semithin cryosections of anthers. Anti-snRNP immunofluorescence. Nuclei of both microspores and tapetal cells (thick arrows) exhibit fluorescence. – Fig. 12. At higher magnification the immunofluorescent nuclei show a bright reticulum with dark regions, while the nucleoli appear negative (thin arrows). – Fig. 13. Nuclear region of a tapetal cell. Formaldehyde fixation, Lowicryl embedding and anti-m3G immunogold labelling; 10 nm gold particles, EDTA staining. In the interchromatin region (*IR*) the EDTA positive fibres (arrows) appear labelled while the granules (arrowheads) are devoid of labelling. The chromatin patches (*Chr*) are unlabelled too. – Bars: 10 μm

Fig. 14–17. Tapetal cells of *Scilla peruviana* (14, 15) and *Capsicum annuum* (16, 17). Early vacuolated stage. – Fig. 14. Semithin cryosection. J26 immunofluorescence. J26 antibody against two nucleolar proteins provides positive fluorescence of the nucleoli (arrows) and the cytoplasm (*Cyt*) of tapetal cells (*Tp*). – Fig. 15. Semithin cryosection. Anti-B36 immunofluorescence. The anti-B36 antibody gives bright fluorescence to the nucleoli (arrows). – Figs. 16, 17. Nuclear regions. Formaldehyde fixation, Lowicryl embedding, immunogold labelling with 10 nm gold particles and EDTA staining. – Fig. 16. Anti-fibrillarin. A specific labelling is exclusively observed over the dense fibrillar component (*F*) of the nucleolus (*Nu*). – Fig. 17. Anti-m3G. The dense fibrillar component (*F*) of the nucleolus (*Nu*) appears labelled. The interchromatin fibres (thin arrows) also show gold particles while the granules (arrowheads) are devoid of labelling as well as the condensed chromatin patches (*Chr*). Fibrillar centres (arrows) and interchromatin region (*IR*). – Bars: Figs. 14, 15: 10 μm, Figs. 16, 17: 1 μm

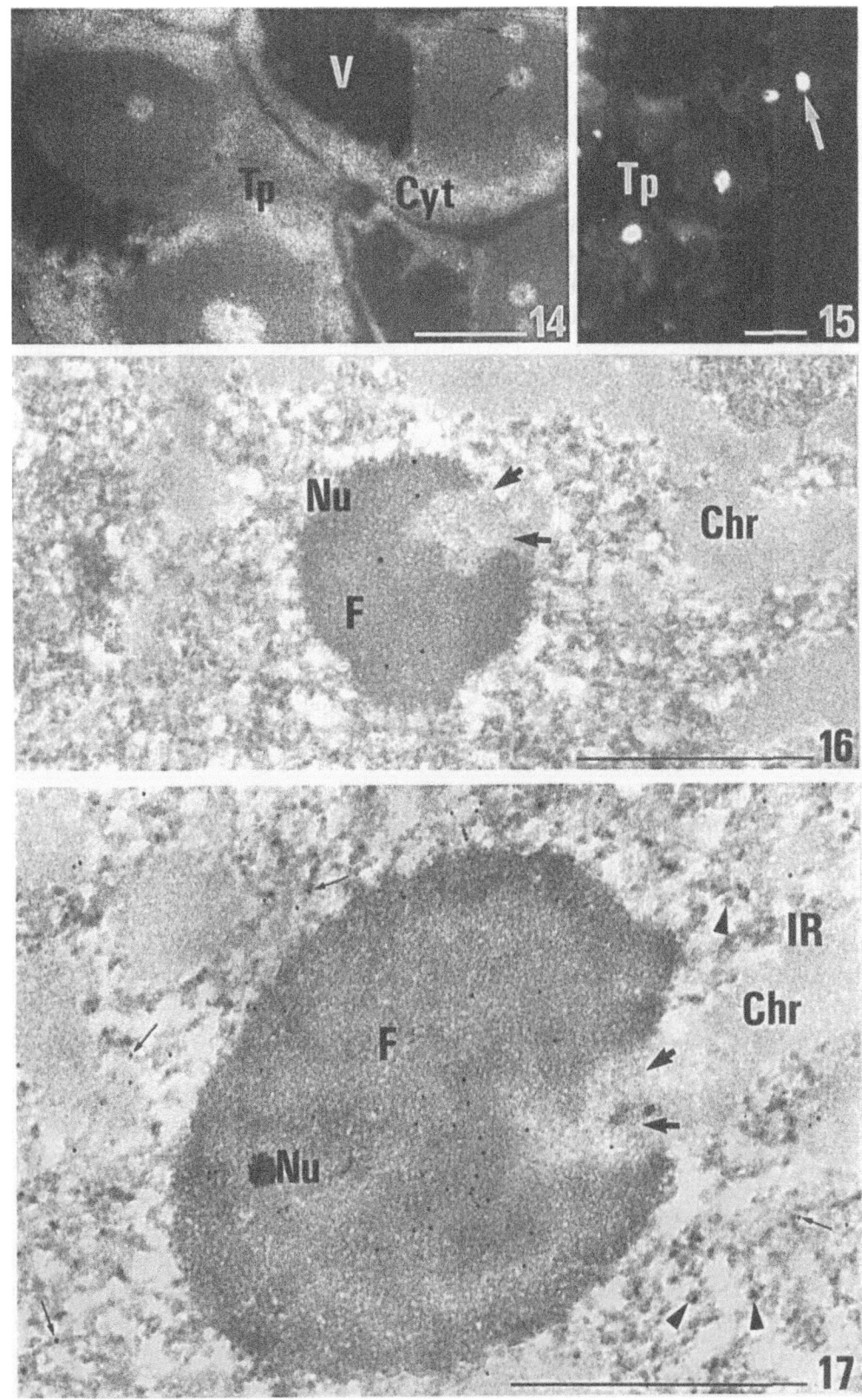

The results with EDTA for RNPs show an interchromatin region rich in granules similar to the perichromatin ones of mammalian cells which are related to forms of storage or transport of RNAs to the cytoplasm (Puvion-Dutilleul & Puvion 1981). Accumulation of these RNP granules has been reported in microspores at distinct stages of development (Testillano & Risueño 1988) and in nuclei with very low or arrested transcriptional activity (Gimenez-Martin & al. 1969, Puvion-Dutilleul & Puvion 1981, Risueño 1990).

The small nuclear ribonucleoprotein particles (snRNPs) contain proteins and snRNAs, and are involved in the splicing, or elimination of introns from the newly synthesized RNAs to form mature messenger RNAs (Steitz 1988). There are also some snRNPs in the nucleolus (snoRNPs) participating in the processing of ribosomal RNAs (Tyc & Steitz 1989). All these snRNAs have the common feature of the trimethyl-guanosine (m3G) in their cap, which enables their immunolocalization by using anti-m3G antibodies (Lührmann & al. 1982). Specific antibodies against common proteins of nucleoplasmic snRNPs (Lührmann & al. 1990) or against nucleolar proteins belonging to the snoRNPs (Tyc & Steitz 1989) localize these other elements of the particles and enable the differential localization of both types of snRNPs: the nucleoplasmic and the nucleolar ones.

Using such antibodies for immunofluorescence and immunogold labelling, the snRNPs are localized on the EDTA-positive fibrillar structures of the interchromatin region of the tapetal nuclei. The distribution of the snRNPs in plant nuclei and their localization in IR fibres has been shown in a few papers dealing with proliferative plant cells (Sanchez-Pina & al. 1989, 1990; Testillano & al. 1991a, 1993; Vazquez-Nin & al. 1992). Therefore, the RNP interchromatin fibrillar structures which are stained by EDTA represent sites of splicing processes in these tapetal nuclei.

The nuclear bodies observed in the interchromatin region of the tapetal nuclei show cytochemical characteristics similar to the dense bodies described in root meristematic cells (Risueño & al. 1978, Risueño & Medina 1986). They are mainly formed by RNPs and no DNA is detected with both cytochemical and immunocytochemical methods. The fact that the localization with anti-snRNPs and anti-m3G is negative in these nuclear bodies could indicate that they are not involved in the splicing reaction and probably take part in other nuclear functions such as the processing, storage, and/or transport of mRNAs and rRNAs, as postulated for other plant cells (Risueño & al. 1978, Risueño & Medina 1986).

The nucleoli of the tapetum show typical morphologies of lowly active and inactive nucleoli, i.e. segregated and compact respectively, but with a large heterogeneous fibrillar centre connected with the chromosome bearing the nucleolar organizer region. The segregated organization, with the granular component surrounding the dense fibrillar one is mainly found in the tapetum at early microsporogenesis stages, probably corresponding to a high decrease in the ribosomal biogenesis rate (Medina & al. 1983a, b; Risueño & Medina 1986). The heterogenous fibrillar centres are characteristic of plant nucleoli with low activity (Risueño & al. 1982, Medina & al. 1983b), and the absence of granular component also corresponds to nucleoli with low transcriptional activity (Risueño & Medina 1986, Risueño & al. 1988).

The anti-m3G antibody also localizes snoRNAs in the nucleolus of the tapetal cells, concretely on the dense fibrillar component, but not on the fibrillar centres. This is consistent with the detection of either fibrillarin or B36 proteins which also appear on the dense fibrillar component where early rRNA processing takes place (JORDAN 1991). Both antigens represent homologous proteins associated with the snoRNAs (TYC & STEITZ 1989). The presence of these proteins has been reported for the dense fibrillar component of many other cells and in several plant cells, independently of the transcriptional stage of activity (TESTILLANO & al. 1992) but no data were available on the immunolocalization with anti-m3G on plant cells, as far as we know.

The immunolocalization of J26 antigen in the nucleolus and the cytoplasm of the tapetal cells indicates the presence of proteins involved in some steps of rRNA processing, which probably are transported with the mature ribosomes to the cytoplasm in these cells. Similar results have been reported in other plant cell types although the precise function of this antigen is still unknown (SANCHEZ-PINA & al. 1989, 1990).

All these results indicate the conservation of important molecules playing key roles in the nuclear compartments of highly specialized plant tissues, such as the tapetum. Localizing different nuclear proteins, RNA, and DNA in the tapetum during pollen ontogeny gives new insights in the understanding of the complex process of pollen formation.

The authors wish to thank Dr E. PUVION and Dr F. PUVION-DUTILLEUL (Villejuif) for the facilities given at their laboratory in which a part of this work was done during a stay of P.S.T. supported by the Spanish Ministry of Education; Dr S. MÜLLER (Strasbourg), Dr J. P. FUCHS (Strasbourg), Dr M. CHRISTENSEN (Kansas), Dr RAMAEKERS (Wageningen) and Dr R. LÜHRMANN (Marburg) for kindly providing us with the anti-H2B, anti-snRNPs, anti-B36, J26, and anti-m3G antibodies respectively. Thanks are also due to Ms BERYL WALKER for checking the English style, Mr JOSÉ BLANCO for the photographic work and Ms OLVIDO PARTEARROYO for typing the manuscript. This work was supported by project DIGICYT/CSIC 88/92 PB033201.

References

ALBERTINI, L., SOUVRÉ, A., AUDRAN, J. D., 1987: Le tapis de l'anthère et ses relations avec les microsporocytes et les grains de pollen. – Rev. Cytol. Biol. végét. – Botan. **10**: 211–242.

ALCHE, J. D., ROMERO, A. T., GARRIDO, M., RODRIGUEZ-GARCIA, M. I., 1990: Factors affecting samples processing for T.E.M. during microsporogenesis in *Olea europaea*. – Proc. MICRO 90, London, 2-6 July 1990, Chapter 17, pp. 627–630. – London: IOP Publishing.

ARIS, J. P., BLOBEL, G., 1988: Identification and characterization of a yeast nucleolar protein that is similar to a rat liver nucleolar protein. – J. Cell. Biol. **107**: 17–31.

BERNHARD, W., 1969: A new staining procedure for electron microscopical cytology. – J. Ultrastruct. Res. **27**: 250–265.

BHANDARI, N. N., 1984: The Microsporangium. – In JOHRI, B. M., (Ed.): Embryology of angiosperms, pp. 53–122. – Berlin, Heidelberg, New York: Springer.

BUSS, P. A., LERSTEN, N. R., 1975: Survey of tapetal nuclear number as a taxonomic character in *Leguminosae*. – Bot. Gaz. **136**: 388–395.

CARNIEL, K., 1963: Das Antherentapetum. Ein kritischer Überblick. – Österr. Bot. Z. **110**: 145–176.

CHAPMAN, G. P., 1967: The tapetum. – Int. Rev. Cytol **107**: 111–126.

CHRISTENSEN, M. E., MOLOO, J., SWISCHUK, J. L., SCHELLING, M. E., 1986: Characterization of the nucleolar protein, B-36, using monoclonal antibodies. – Exp. Cell. Res. **166**: 77–93.

FAKAN, S., PUVION, E., 1980: The ultrastructural visualization of nucleolar and extranucleolar RNA synthesis and distribution. – Int. Rev. Cytol. **65**: 255–299.

FITZGERALD, M., 1993: Secretory events in the freeze-substituted tapetum of the orchid *Pterostylis concinna*. – Pl. Syst. Evol. [Suppl.] **7**: 53–62.

GIMENEZ-MARTIN, G., RISUEÑO, M. C., LOPEZ-SAEZ, J. E., 1969: Generative cell envelope in pollen grains as a secretion system: a postulate. – Protoplasma **67**: 233–235.

HARRIS, J. R., ZBARSKY, I. B., (Eds) 1990: Nuclear structure and function. – New York, London: Plenum Press.

HESSE, M., HESS, M. W., 1993: Recent trends in tapetum research. A cytological and methodological review. – Pl. Syst. Evol. [Suppl.] **7**: 127–145.

HORNER, H. T. JR, ROGERS, M. A., 1974: A comparative light and electron microscopic study of microsporogenesis in male-fertile and cytoplasmic male-sterile pepper (*Capsicum annuum*). – Canad. J. Bot. **52**: 435–449.

JACKSON, D. A., 1991: Structure-function relationships in eukaryotic nuclei. – Bio Essays **13**: 1–10.

JORDAN, E. G., 1991: Interpreting nucleolar structure: where are the transcribing genes? – J. Cell Sci. **98**: 437–442.

KEIJZER, C. J., 1987: The processes of anther dehiscence and pollen dispersal. I. The opening mechanism of longitudinally dehiscing anthers. – New Phytol. **105**: 487–498.

LÜHRMANN, R., APPEL, B., BRINGMANN, P., RINKE, J., REUTER, R., ROTHE, S., 1982: Isolation and characterization of rabbit anti-m$_3$ 2,2,7 G antibodies. – Nucleic Acids Res. **10**: 7103–7113.

– KASTNER, B., BACH, M., 1990: Structure of spliceosomal snRNPs and their role in pre-mRNA splicing. – Biochim. Biophys. Acta **1087**: 265–292.

LUTZ, Y., 1986: Caracterisation de protéines du reseau RNP nucleaire de cellules HeLa à l'aide d'anticorps monoclonaux. – Thèse doctorale. Univ. Louis Pasteur, Paris.

MAJEWSKA-SAWKA, A., JASSEM, B., MACEWICZ, J., RODRIGUEZ-GARCIA, M. I., 1990: An electron microscopic study of anther structure in male fertile and male-sterile sugar beets: tapetum development. – In BLANCA, G. & al. (Eds): Polen, esporas y sus aplicaciones. pp. 57–63. – Granada: University of Granada.

MEDINA, F. J., RISUEÑO, M. C., RODRIGUEZ-GARCIA, M. I., SANCHEZ-PINA, M. A., 1983a: The NOR and fibrillar centers during plant gametogenesis. – J. Ultrastruct. Res. **85**: 300–310.

– MORENO DIAZ DE LA ESPINA, S., 1983b: 3-D Reconstruction and morphometry of fibrillar centres in plant cells in relation to nucleolar activity. – Biol. Cell. **48**: 31–38.

MÜLLER, S., CHAIX, M. L., BRIAND, J. P., VAN REGENMORTEL, M. H., 1991: Immunogenicity of free histones and of histones complexed with RNA. – Molec. Immunol. **28**: 763–772.

PUVION-DUTILLEUL, F., PUVION, E., 1981: Relationship between chromatin and perichromatin granules in cadmium-treated isolated hepatocytes. – J. Ultrastruct. Res. **74**: 341–350.

RISUEÑO, M. C., 1990: Pollen biology: Structure and function. – In BLANCA, G., & al. (Eds): Polen, esporas y sus aplicaciones, pp. 31–42. – Granada: University of Granada.

– MORENO DIAZ DE LA ESPINA, S., 1979: Ultrastructural and cytochemical study of the quiescent root meristematic cell nucleus. – J. Submicr. Cytol. **11**: 85–95.

– MEDINA, F. J., 1986: The nucleolar structure in plant cells. – Rev. Biol. Cell. **7**: 1–163.

- Mᴏʀᴇɴᴏ Dɪᴀᴢ ᴅᴇ ʟᴀ Esᴘɪɴᴀ, S., Fᴇʀɴᴀɴᴅᴇᴢ Gᴏᴍᴇᴢ, M. E., Gɪᴍᴇɴᴇᴢ-Mᴀʀᴛɪɴ, G., 1978: Nuclear micropuffs in *Allium cepa* cells. I. Quantitative, ultrastructural and cytochemical study. – Cytobiologie **16**: 209–223.
- Mᴇᴅɪɴᴀ, F. J., Mᴏʀᴇɴᴏ Dɪᴀᴢ Dᴇ Lᴀ Esᴘɪɴᴀ, S., 1982: Nucleolar fibrillar centres in plant meristematic cells: ultrastructure, cytochemistry and autoradiography. – J. Cell Sci. **58**: 313–329.
- Tᴇsᴛɪʟʟᴀɴᴏ, P. S., Sᴀɴᴄʜᴇᴢ-Pɪɴᴀ, M. A., 1988: Variations of nucleolar ultrastructure in relation to transcriptional activity during G_1, S and G_2 periods of microspore interphase. – In Cʀᴇsᴛɪ, M., Pᴀᴄɪɴɪ, E., (Eds): Sexual reproduction in higher plants, pp. 9–14. – Berlin: Springer.
- Sᴀɴᴄʜᴇᴢ-Pɪɴᴀ, M. A., Kɪᴇғᴛ, H., Sᴄʜᴇʟ, J. H. N., 1989: Immunocytochemical detection of non-histone nuclear antigens in cryosections of developing somatic embryos from *Daucus carota* L. – J. Cell Sci. **93**: 615–622.
- Kɪᴇғᴛ, H., Sᴄʜᴇʟ, J. H. N., Tᴇsᴛɪʟʟᴀɴᴏ, P. S., Rɪsᴜᴇñᴏ, M. C., 1990: Localization of non-histone nuclear proteins by immunocytochemistry in somatic embryos and pollen grains. – In Hᴀʀʀɪs, J. R., Zʙᴀʀsᴋʏ, I. B., (Eds): Nuclear structure and function, pp. 253–258. – New York: Plenum Press.
- Sʜɪᴠᴀɴɴᴀ, K. R., Jᴏʜʀɪ, B. M., 1985: The angiosperm pollen, structure and function. – New Delhi, Bangalore, Bombay: Wiley Eastern Limited.
- Sᴛᴇɪɴʙʀᴇᴄʜᴛ, R. A., Zɪᴇʀᴏʟᴅ, K., (Eds) 1987: Cryotechniques in biological electron microscopy. – Berlin, Heidelberg, New York: Springer.
- Sᴛᴇɪᴛᴢ, J. A., 1988: "Snurps". – Scientific American. **June 1988**: 36–41.
- Sᴛᴏᴄᴋᴇʀᴛ, J. C., Dᴇʟ Cᴀsᴛɪʟʟᴏ, P., Tᴇsᴛɪʟʟᴀɴᴏ, P. S., Rɪsᴜᴇñᴏ, M. C., 1989: Notes on technics. Fluorescence of plastic embedded tissue sections after curcumin staining. – Stain Technol. **64**: 207–209.
- Tᴇsᴛɪʟʟᴀɴᴏ, P. S., Rɪsᴜᴇñᴏ, M. C., 1988: Evolution of nuclear interchromatin structures during microspore interphase periods. – In Cʀᴇsᴛɪ, M., Pᴀᴄɪɴɪ, E., (Eds): Sexual reproduction in higher plants, pp. 151–156. –Berlin: Springer.
- Oʟᴍᴇᴅɪʟʟᴀ, A., Sᴀɴᴄʜᴇᴢ-Pɪɴᴀ, M. A., Rɪsᴜᴇñᴏ, M. C., 1991a: Detection of small nuclear RNAs in plant nuclei. – In Abstracts of the Colloque Franco-Iberique de M. E. pp. 38–39. – Barcelona.
- Sᴀɴᴄʜᴇᴢ-Pɪɴᴀ, M. A., Oʟᴍᴇᴅɪʟʟᴀ, A., OʟʟᴀᴄᴀʀɪᴢQᴜᴇᴛᴀ, M. A., Tᴀɴᴅʟᴇʀ, C. J., Rɪsᴜᴇñᴏ, M. C., 1991b: A specific ultrastructural method to reveal DNA: The NAMA-Ur. – J. Histochem. Cytochem. **39**: 1427–1438.
- – Lᴏᴘᴇᴢ-Iɢʟᴇsɪᴀs, C., Oʟᴍᴇᴅɪʟʟᴀ, A., Cʜʀɪsᴛᴇɴsᴇɴ, M. E., Rɪsᴜᴇñᴏ, M. C., 1992a: Distribution of B-36 nucleolar protein in relation to transcriptional activity in plant cells. – Chromosoma **102**: 41–49.
- Oʟᴍᴇᴅɪʟʟᴀ, A., Sᴀɴᴄʜᴇᴢ-Pɪɴᴀ, M. A., Rᴀsᴋᴀ, I., Rɪsᴜᴇñᴏ, M. C., 1992b: The use of ultrathin cryosections for immunoelectron microscopy in plant cell nucleus. – Electron Microscopy 3: EUREM 92, pp. 87–88. – Granada, Spain.
- Sᴀɴᴄʜᴇᴢ-Pɪɴᴀ, M. A., Oʟᴍᴇᴅɪʟʟᴀ, A., Fᴜᴄʜs, J. P., Rɪsᴜᴇñᴏ, M. C., 1993: Characterization of the interchromatin region as the nuclear domain containing snRNPs in plant cells: a cytochemical and immunoelectron microscopy study. – European J. Cell Biol. **61**(2) (in press).
- Tʏᴄ, K., Sᴛᴇɪᴛᴢ, J. A., 1989: U3, U8 and U13 comprise a new class of mammalian snRNPs localized in the cell nucleolus. – EMBO J. **8**: 3113–3119.
- VᴀᴢQᴜᴇᴢ-Nɪɴ, G. H., Eᴄʜᴇᴠᴇʀʀɪᴀ, O. M., Mɪɴɢᴜᴇᴢ, A., Mᴏʀᴇɴᴏ Dɪᴀᴢ Dᴇ Lᴀ Esᴘɪɴᴀ, S., Fᴀᴋᴀɴ, S., Mᴀʀᴛɪɴ, T. E., 1992: Ribonucleoprotein components of root meristematic cell nuclei of the tomato characterized by application of mild loosening and immunocytochemistry. – Exp. Cell. Res. **200**: 431–438.

Verheijen, R., Kuijpers, H., Vooijs, P., Van Venrooij, W., Ramaekers, F., 1986: Protein composition of nuclear matrix preparations from HeLa cells: an immunochemical approach. – J. Cell. **80**: 103–122.

Addresses of the authors: Pilar S. Testillano[1,2], Pablo Gonzalez-Melendi[1], Begoña Fadon[1], Amelia Sanchez-Pina[1,α], Adela Olmedilla[1,β], Maria Del Carmen Risueño[1] (correspondence) – [1] Centro de Investigaciones Biológicas, CSIC, Velázquez 144, E-28006 Madrid, Spain. – [2] Departamento CC. Morfológicas y Cirugía, Facultad Medicina, Universidad Alcalá de Henares, Madrid. – [α] Present address: C.E.B.A.S., CSIC, Murcia, Spain. – [β] Present address: Est. Exp. del Zaidín, CSIC, Granada, Spain.

Pl. Syst. Evol. [Suppl.] 7: 91–97 (1993)

Nuclease activities in *Tradescantia paludosa* and *Brassica napus* pollen and tapetum

M.-Françoise Jardinaud, André Souvré, and Gilbert Alibert

Key words: *Commelinaceae, Tradescantia paludosa, Brassicaceae, Brassica napus.* – Male gametophyte, tapetum, nuclease.

Abstract: In several angiosperm species, nuclease activities of tapetal origin were localized in the outer part of the pollen grain wall. These hydrolytic enzymes are an obstacle to direct gene transfer into pollen and to androgenetic transformation. The presence and the activity of nucleases in the male gametophyte and in the tapetum of *Tradescantia paludosa* (periplasmodial tapetum) and of *Brassica napus* (secretory tapetum) were assessed.

Nucleases were active in *Tradescantia paludosa* male gametophyte and in the corresponding tapetum from the tetrad stage to the young pollen grain stage. Addition of 1 mM spermidine blocks nuclease activity in the pollen, but not in the tapetum.

Nuclease activities were also found in the pollen of *Brassica napus* and in the surrounding locular fluid but only in aged microspores and young pollen grains. Washings removed most of the nucleases from the microspores, but only partly from the pollen grains.

The use of pollen as a gene vector is a highly valuable methodology in plant genetic transformation since it can allow in vitro fertilization after maturation of fertile pollen grains, or in vitro regeneration of doubled haploid transgenic plants.

Direct gene transfer into the male gametophyte has not given convincing results, except for the transformation of tobacco pollen grains by bombardment (Twell & al. 1989). One of the possible reasons for the lack of positive results is the presence of pollen nucleases, which can hydrolyze the transformant naked DNA before its introduction into the pollen grain cytoplasm (Van Der Westhuisen & al. 1987, Roeckel & al. 1988). Nucleases were supposed to be only of tapetal origin (Shivanna & Johri 1985) or partly of gametophytic origin (Matousek & Tupy 1987). The proteins released by the tapetum during the late stages of microsporogenesis until pollen grain maturation are incorporated into the exine of the microspores or pollen grains. They are also associated with lipid compounds to form the tryphine of the pollen grain coat (Albertini & al. 1987). These proteins include enzymes but also proteins involved is sporophytic self-incompatibility and as well as allergens (Shivanna & Johri 1985).

Two tapetum types have been described: in most angiosperm species, one layer of tapetal cells (termed cellular, parietal or secretory tapetum), maintains its cellular structure until the late developmental stages of pollen. After the disorganisation of the cellular structure, microspores or pollen grains float in a locular fluid formed by the content of the tapetal cells. The second and less frequent type is the (peri-)plasmodial tapetum. In *Tradescantia*, the plasmodial structure develops at the beginning of meiotic prophase I, then the plasmodium invades the anther loculus and surrounds microspore mother cells, tetrads, microspores, and pollen grains successively. Then it degenerates.

Up to now investigations have concerned only the nuclease activities of the pollen in species with a secretory tapetum. As far as we know, no investigations were undertaken for the secretory tapetum itself and for pollen and tapetum in the species of the plasmodial type.

As a prerequisite to the transformation of the male gametophyte of *Brassica napus*, the change in pollen and in locular fluid nuclease activities from the tetrad stage to the young pollen grains was investigated in the present study. Similar investigations were carried out on *Tradescantia paludosa*, a plant with a plasmodial tapetum, which was used as an experimental model. In both, the possibilities of inhibiting the nuclease activities were also studied.

Material and methods

Pollen and tapetum extraction. *Brassica napus L.* var. *oleifera* DC. cv. 'Tapidor' and *Tradescantia paludosa* L. plants were grown on experimental plots and in a greenhouse, respectively.

The procedures used for pollen and locular fluid extraction were the same as those used for microspore culture (Huang & al. 1990). For *Tradescantia paludosa* and *Brassica napus* pollen and tapetum extraction, N6 (Dunwell 1985) and B5K medium (Huang & al. 1990) were used, respectively.

Anthers of *Tradescantia paludosa*, sterilisated with sodium hypochlorite, were grounded with a pestle in N_6 medium. In contrast, flower buds of *Brassica napus*, sterilisated likewise, were crushed with a Waring blender in B_5K medium. All material was then filtered through a $50\,\mu m$ steel filter and centrifugated for 5 min at 60 g. The pellet held intact microspores or pollen grains, while the respective supernatant consisted either to material of the *Tradescantia paludosa* plasmodial tapetum or to the locular fluid of *Brassica napus*.

The concentration of pollen was adjusted to 10^6 cells/ml. The medium used to test the nuclease activities was a modified CS medium (Clark & Steer 1983) with 10% sucrose and pH 6.7 for *Tradescantia paludosa* and with 13% sucrose and pH 5.9 for *Brassica napus* (Jardinaud & al. 1993).

Nuclease activities. For the study of the diffusible nuclease activities, first 25000 each of tetrads, microspores or pollen grains, then the tapetal content of all the anthers used for the extractions were incubated with plasmid Zmg 13 which is pollen specific (Hamilton & al. 1989) at a final concentration of $25\,\mu g/ml$ in modified CS medium.

The mixtures of DNA and pollen, or DNA and corresponding tapetum were incubated at $30\,^{\circ}C$ during 5 to 60 min. After centrifugation (60 g, 5 min), $5\,\mu L$ of a 40% sucrose – 0.25% bromophenol blue solution was added to the supernatant which contained the exogenous DNA.

The supernatant DNA solution was immediately electrophoretically tested (100 V, 90 min) on agarose gel (40 mM tri-acetate, 1 mM EDTA, 2 mg/ml ethidium bromide BEt, 1% agarose) in TAE buffer (40 mM tri-acetate, 1 mM EDTA). After electrophoresis the fluorescence of the DNA banding on the gel was analysed under U.V. light; the image banding definition was improved using Image Analysis System (Biocom 500, Program Imagenia, Les Ulis, France).

To inhibit the nuclease activities of both species, tetrads, microspores or pollen grains were first washed three times with CS medium, then, if necessary, 1 mM spermidine and the plasmid were added simultaneously to the CS medium. The electrophoretic analysis was carried out as described above.

Results

Tradescantia paludosa. **During pollen development of** *Tradescantia paludosa* the nuclease activity level was related to the pollen developmental stage and increased with time (Fig. 1). The plasmid was only partially degraded after incubation with microspore mother cells at prophase I and tetrads but completely disappeared in the presence of vacuolated microspores (Fig. 1a). Five minutes of incubation were sufficient to hydrolyse the DNA at the microspore stage but 60 minutes were necessary to reach the same degradation level at the tetrad stage (Fig. 1b).

When tetrads and vacuolated uninucleate microspores were washed three times (Fig. 1c), part of the nuclease activity was removed; nevertheless, the nuclease activity was maintained in young pollen grains. The spermidine treatment inhibited the remaining nuclease activity, irrespective of the developmental stage and of the duration of incubation (Fig. 1d).

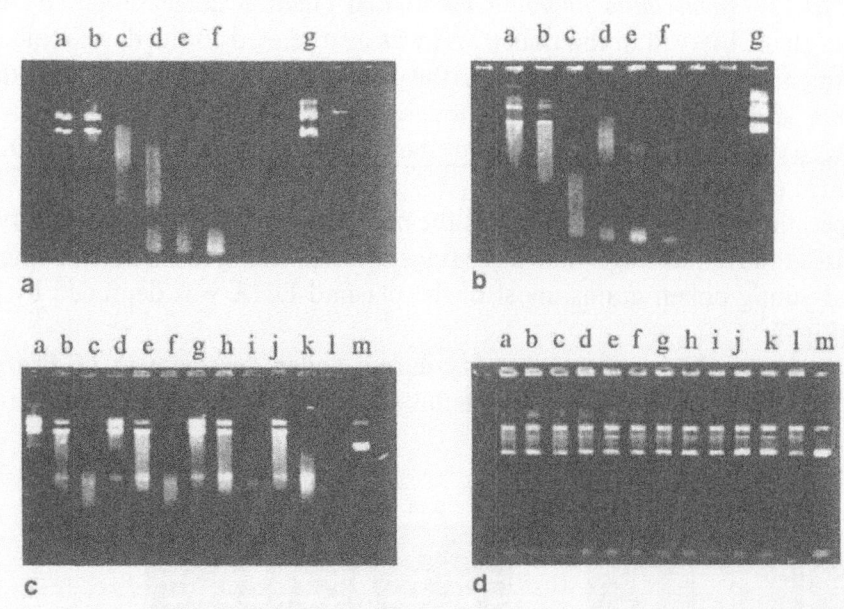

Fig. 1. Nuclease activity of *Tradescantia paludosa* pollen and inhibition assays. Nuclease activity at different stages of microspore pollen development: microspore mother cells (lane a), tetrads (lane b), non vacuolated microspores (lane c), vacuolated microspores (lane d), young pollen grains (lane e) and mature pollen grains (lane f), after 5 (panel *a*) or 60 minutes (panel *b*) of incubation. Effects of washing (panel *c* or addition of 1 mM spermidine (panel *d*) on the inhibition of the nuclease activity at different developmental stages of pollen: tetrads (laines a, d, g, j), vacuolated uninucleate microspores (lanes b, e, h, k) and young pollen grains (lanes c, f, i, l) for different incubation times: 5 (lanes a, b, c), 10 (lanes d, e, f), 20 (lanes g, h, i), and 60 (lanes j, k, l) minutes. Lanes g, m: control (Zmg 13 plasmid)

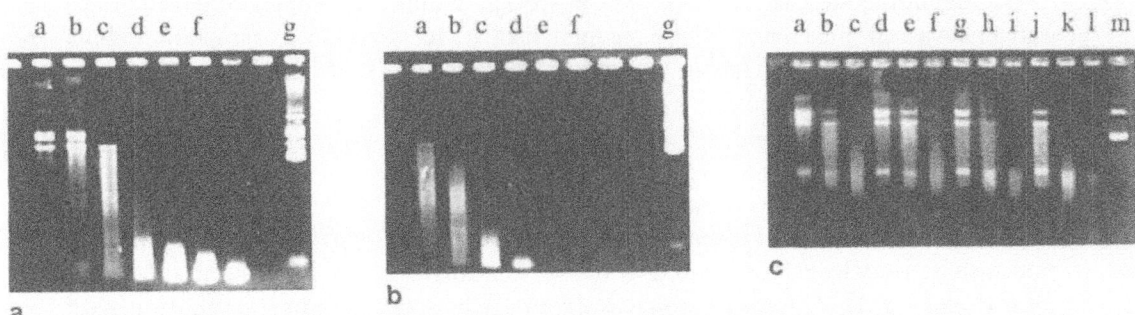

Fig. 2. Nuclease activity of *Tradescantia paludosa* tapetum and inhibition assays. Nuclease activity at different developmental stages of tapetum corresponding to: microspore mother cells (lane a), tetrads (lane b), non-vacuolated microspores (lane c), vacuolated microspores (lane d), young pollen grains (lane e), and mature pollen grains (lane f), after 5 (panel a) or 60 minutes (panel b) of incubation. Effects of 1 mM spermidine addition on the inhibition (panel c) of the nuclease activity at different developmental stages of tapetum corresponding to: tetrads (lanes a, d, g, j), vacuolated uninucleate microspores (lanes b, e, h, k) and young pollen grains (lanes c, f, i, l) for different incubation times: 5 (lanes a, b, c), 10 (lanes d, e, f) 20 (lanes g, h, i), and 60 (lanes j, k, l) minutes. Lanes g, m: control (Zmg 13 plasmid)

The level of ***Tradescantia paludosa* plasmodial tapetum** nuclease activity was higher than that observed in the pollen and was also related to the developmental stage. During the young microspore stage, the plasmid was completely degraded by the tapetum after only five minutes of incubation (Fig. 2a); after 60 minutes of incubation, at each developmental stage of the tapetum, most of the plasmid DNA was degraded (Fig. 2b).

The spermidine treatment suppressed the major part of the nuclease activity of the tapetum at the tetrad stage, but, at the stage corresponding to vacuolated microspores and young pollen grains, most of the plasmid DNA was degraded by the tapetum (Fig. 2c).

Brassica napus*.** The nuclease activity **during pollen development of *Brassica napus was also related to the developmental stage (Fig. 3a). No nuclease activity

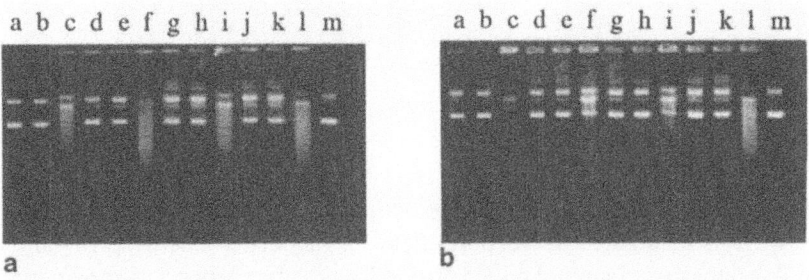

Fig. 3. *Brassica napus* pollen nuclease activity and inhibition assays. Nuclease activities (panel *a*) and inhibition after three washings (panel *b*) at three different development stages of microspore pollen: tetrads (a, d, g, j), uninucleate microspores (b, e, h, k), and young pollen grains (c, f, i, l) for different incubation times: 5 minutes (a, b, c), 10 minutes (d, e, f), 20 minutes (g, h, i) and 60 minutes (j, k, l). lane m: control (Zmg 13 plasmid).

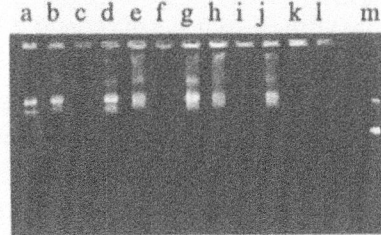

Fig. 4. Analysis of *Brassica napus* tapetum nuclease activities. Nuclease activity of tapetum corresponding to three different stages of microspore pollen development: tetrads (a, d, g, j), uninucleate microspores (b, e, h, k) and young pollen grains (c, f, i, l) for different incubation times: 5 minutes (a, b, c), 10 minutes (d, e, f), 20 minutes (g, h, i) and 60 minutes (j, k, l). lane m: control (Zmg 13 plasmid).

was observed in tetrads and microspores, irrespective of the incubation time (from 0 to 60 min). However, high nuclease activities were observed in the pollen grains after only five minutes of incubation. Three washings with the medium decreased the nuclease activity of the pollen but a low level of activity was maintained (Fig. 3b).

The nuclease activity of **Brassica napus locular fluid** was also related to the stage of development. No nuclease activity was found at the tetrad stage but it appeared in microspores, and a high level of nuclease activity was measured in young pollen grains (Fig. 4). Plasmid DNA degradation increased with the incubation time. The inhibition of nuclease activity in the locular fluid of rape was not investigated.

Discussion

The nuclease activities observed in *Tradescantia paludosa* and *Brassica napus* pollen for the previously determined electroporation conditions (JARDINAUD & al. 1993) confirm the results of MATOUSEK & TUPY (1985). Observation of DNA degradation at early developmental stages suggested that nuclease activity was already present in *Tradescantia paludosa* pollen mother cells but was detected only in young pollen grains in *Brassica napus*. The late nuclease activity in rape might be related to the low level of DNA degradation comparatively to *Zea mays*, *Secale cereale* and *Pinus nigra* (MATOUSEK & TUPY 1985).

The occurrence of pollen nuclease release into the electroporation medium after only five minutes of incubation, proved the external localization of at least a large part of these enzymes on the pollen grain, i.e. in the exine or in the pollen grain coat (ALBERTINI & al. 1987, MATOUSEK & TUPY 1985).

Tradescantia paludosa plasmodial tapetum and *Brassica napus* locular fluid displayed a high nuclease activity at the pollen grain stage, which affirms the sporophytic origin of pollen nucleases (SHIVANNA & JOHRI 1985) in species with a plasmodial tapetum as well as in species with a cellular tapetum.

The presence of high levels of nuclease activities in the first developmental stages after the meiosis of pollen mother cells in *Tradescantia paludosa*, and only from the young microspore stage in rape, suggested that the enzymatic activity is likely to be related to tapetum physiology. Plasmodial tapetum formation in *Tradescantia paludosa* begins indeed at the meiotic prophase I, with the lysis of the cell walls of the tapetal layer already present at premeiosis (MEPHAM & LANE 1969) while the disorganisation of *Brassica* tapetum cellular structure with the release of tapetal cell cytoplasm in the loculus occurs much later (PACINI & al. 1985). It seems that a relation exists between tapetum structure disorganisation, which results in a close contact of the tapetal cytoplasm with the pollen, and the tapetum ability to synthesize hydrolytic enzymes which are then integrated into the pollen grain wall.

In order to prevent degradation of exogenous DNA in transformation experiments, several methods have been put forward for the inhibition of nuclease activity: successive washings (ROECKEL & al. 1988), addition of EDTA (NEGRUTIU & al. 1987), PEG (ROECKEL & al. 1988) or spermidine (GALSTON & al. 1979). Our investigations, show that: (i) three washings were enough to remove most of the *Brassica napus* pollen or microspore nuclease activity; (ii) washings were inefficient in the case of *Tradescantia paludosa* pollen; (iii) 1 mM spermidine thoroughly inhibited the nuclease activities at all developmental stages of pollen but the activity of tapetum nucleases was only slightly decreased by this treatment.

The knowledge of the change in pollen nuclease activity during pollen formation is a prerequisite to the use of the electroporation technique for the genetic transformation of the microspores of rape, at a developmental stage where pollen nuclease activity is low and easy to remove.

These results were used for the development of the first transient expression experiments of GUS gene (JARDINAUD & al. 1993) under control of Cam 35S promoter (plasmid pCGN1: VASSER, Wageningen, The Netherlands) and Zmg 13 promoter (plasmid Zmg 13: MASCARENHAS, Albany, USA). Further investigations are being carried out to check whether transient expression will be further affirmed by stable expression in rape plantlets regenerated from microspores.

The authors are grateful to Prof. J. P. MASCARENHAS (State University of New York, Albany, N.Y.) for the supply of plasmid Zmg 13.

References

ALBERTINI, L., SOUVRÉ, A., AUDRAN, J. C., 1987: Le tapis de l'anthère et ses relations avec les microsporocytes et les grains de pollen. – Rev. Cytol. Biol. Végét. Bot. **10**: 211–242.

CLARK, E., STEER, M. W., 1983: Cytoplasmic structure of germinated and ungerminated pollen grains of *Tradescantia paludosa*. – Caryologia **36**: 299–305.

DUNWELL, J. M., 1985: Haploid cell culture. – In DIXON, R. A., (Ed.): Plant cell culture: a practical approach. – Oxford, Washington: IRL Press Limited.

GALSTON, A. W., SAWHNEY, R. K., ALTMAN, A., FLORES, H., 1979: Polyamines, macromolecular syntheses and the problem of cereal regeneration. – In FERENCZY, L., FARKAS, G.L., (Eds): Advances in protoplast research. Proceedings of the 5th Int. Protoplast Symp., Szeged, Hungary 1979, pp. 485–497. – New York: Pergamon Press.

HAMILTON, D. A., BASHE, D., STINSON, J. R., MASCARENHAS, J. P., 1989: Characterisation of a pollen specific genomic clone of maize – Sex. Pl. Reprod. **2**: 208–212.

HUANG, B., BIRD, S., KEMBLE, R., SIMMONDS, D., KELLER, W., MIKI, B., 1990: Effects of culture density, conditioned medium and feeder cultures on microspore embryogenesis in *Brassica napus* L. cv. 'Topas.' – Pl. Cell Reprod. **8**: 594–597.

JARDINAUD, M.-F., SOUVRÉ, A., ALIBERT, G., 1993: Transient Gus gene expression in *Brassica napus* electroporated microspores. – Plant Science (in press).

MATOUSEK, J., TUPY, J., 1985: The release and some properties of nucleases from various pollen species. – J. Pl. Physiol. **119**: 169–178.

– – 1987: Developmental changes in nuclease and other phosphohydrolase activities in anthers of *Nicotiana tabacum* L. – J. Pl. Physiol. **129**: 351–362.

MEPHAM, R. H., LANE, G. R., 1969: Formation of the tapetal periplasmodium in *Tradescantia bracteata*. – Protoplasma **68**: 175–192.

NEGRUTIU, I., MOURAS, A., HORTH, M., JACOBS, M., 1987: Direct gene transfer to plants: present development and some future prospectives. – Pl. Physiol. Biochem. **25**: 493–503.

PACINI, E., FRANCHI, G. G., HESSE, M., 1985: The tapetum: its form, function, and possible phylogeny in *Embryophyta*. – Pl. Syst. Evol. **149**: 155–185.

ROECKEL, P., HEIZMANN, P., DUBOIS, M., 1988: Attempts to transform *Zea mays* via pollen grains. Effects of pollen and stigma nuclease activities. – Sex. Pl. Reprod. **1**: 156–163.

SHIVANNA, K. R., JOHRI, B. M., 1985: The angiosperm pollen. Structure and function. – New Delhi: Wiley Eastern Limited.

TWELL, D., KLEIN, T. M., FROMM, M. E., MCCORMICK, S., 1989: Transient expression of chimeric genes delivered into pollen by microprojectile bombardment. – Pl. Physiol. **91**: 1270–1274.

VAN DER WESTHUISEN, A. J., GLIEMEROTH, A. K., WENZEL, W., HESS, D., 1987: Isolation and partial characterisation of an extracellular nuclease from pollen of *Petunia hybrida*. – J. Pl. Physiol. **131**: 373–384.

Address of the authors: M.-F. JARDINAUD, A. SOUVRÉ, G. ALIBERT, Laboratoire de Biotechnologie et Amélioration des Plantes (EA-DRED 832), INP-ENSAT, 145 Av. de Muret, F-31076 Toulouse Cédex, France.

Pl. Syst. Evol. [Suppl.] 7: 99–105 (1993)

Changes in protein electrophoretic patterns during pollen development in *Tradescantia bracteata* plasmodial tapetum and effect of 2,4-D applications

V. Rasolonjatovo, C. Balague, and A. Souvré

Key words: *Commelinaceae, Tradescantia bracteata.* – 2,4-D herbicide, proteins, tapetum, pollen development.

Abstract: The quantitative (Biorad test) and qualitative (SDS-PAGE) variations in total proteins at 4 successive stages of pollen development in *Tradescantia bracteata* tapetum were determined. Then, the effect of 2,4-D herbicide treatments (4000 ppm and 8000 ppm) applied directly to the inflorescences was analyzed. Several modifications were observed in the tapetum until the mature pollen grain stage. The presence of new polypeptid bands affirms the occurrence of protein synthesis at late stages of pollen development. Two polypeptid bands (16 and 34 kDa) were present during all stages and the high molecular weight proteins disappeared from the tapetum at the beginning of pollen maturation. 2,4-D treatment modified the protein patterns of *Tradescantia* tapetum. The 16 and 34 kDa bands were maintained after the treatments, and the appearance of a set of new bands was observed concomitantly with the disappearance of several polypeptidic bands: a 62 kDa protein was induced during all stages by the 8000 ppm treatment, in addition to the 74 kDa (8000 ppm) and 23 kDa (4000 ppm) bands present at the first three stages of pollen development. Some of the new bands were specific of one stage and one treatment. The effect of 2,4-D treatments on the tapetum is discussed in relation to the important role of this tissue in pollen quality.

Mepham & Lane (1969) and Tiwari & Gunning (1986) have described the ultra-structural development of the *Tradescantia* tapetum. In its plasmodial tapetum, the tapetal cell walls are hydrolysed at the onset of meiosis in the sporogenous tissue, and the tapetal cell protoplasts fuse early to form a plasmodium around the microspore mother cells (MMC's) in the anther loculus. The formation of a peri-plasmodium and the associated changes which take place during the tapetal development must be considered as a cell and functional reorganization and not as a degenerative process. Lipid, polysaccharide, and protein synthesis are maintained from the binucleate stage until anthesis. Tapetal cells are known to play a nutritive role towards the microspores and pollen grains (Albertini & al. 1987). A second tapetal cell function is to release the young haploid microspores from the callose wall enclosing the meiotic tetrads through secretion of a callase (Pacini & al. 1985). A third role for the tapetal cells of cellular tapetum is the production of precursors of the pollen wall constituents (Mascarenhas 1990, Bedinger 1992) but, in the case of a plasmodial tapetum, no precursors of the proteins present in the exine could be characterized so far (Mascarenhas 1990).

The plant responses to chlorophenoxy herbicides as 2,4-D are variable. The response at the cellular level depends on the concentration of the herbicide which is controlled by its penetration into the plant, its subsequent transport and differential accumulation, and by the application rate as well as the growth stage (Martin & al. 1989). Some herbicides disrupt mitosis and inhibit microtubule polymerization and/or alter the organization of spindle microtubules (Vaughn & Lehnen 1991).

2,4-D can enhance or inhibit protein synthesis in the plant and participates in the regulation of nucleic acid metabolism. The primary site of action of 2,4-D is the nucleus (Penner & Ashton 1966). The percentage of anomalous meiotic cells is significantly increased; the chromosome abnormalities observed are bridge formation, lagging chromosomes, chromosomal breaks, and sticky chromosomes (Al Najjar & Soliman 1982). At sublethal doses, flowering and seed production are delayed (Hume & Shirriff 1989).

Plant progeny quality is largely related to the pollen quality of the male parent (Mulcahy 1986) which is itself dependent on the activity of the tapetum. Abnormalities or absence of the tapetum are known to induce male sterility (Mian & al. 1974).

The effects of 2,4-D applications on protein expression in the pollen grains and also in the tapetum of *Tradescantia* were found to account for the negative effects of 2,4-D on pollen quality. The plasmodial tapetum of *Tradescantia* is convenient for a separate extraction of the pollen grains and of the tapetum. The present paper reports on the protein content and development in the tapetum of control plants from the microspore to the mature pollen grain stage and on the changes induced by the application of two doses of 2,4-D.

Material and methods

The inflorescences of *Tradescantia bracteata* (*Commelinaceae*) plants grown in a greenhouse (16 h day/8 h night) were excised and immersed in modified New Jersey nutritive medium supplemented (or not) with 2,4-D (commercial product Desormone, Rhone Poulenc Agrochimie, Lyon, France) at 4000 ppm and 8000 ppm.

Pollen and tapetum extraction. 60 to 120 anthers from each of the 4 pollen developmental stages [microspores (M), young pollen (YP), maturing pollen (MP), and mature pollen (P)] were gently crushed (without strong damage for microspores or pollen grain) in nutritive medium. The slurry was filtered (pore size 50 μm); then the filtrate was centrifuged (60 g for 5 min): the tapetum protoplasm corresponded to the supernatant (tapetum solution) and the pollen grains to the pellet.

Protein extraction and electrophoresis. The tapetum protoplasm constituents were dissociated in liquid N_2. The extraction buffer (10 mM TRIS, 5 mM $MgCl2$, 2.3% SDS; pH 6.8) was then added and the homogenate was centrifuged at 15,000 g for 10 min and the supernatant was stored at $-20°C$. The total protein concentration in the solution was determined spectrophotometrically using the Biorad test. One-dimensional SDS-PAGE was carried out according to the procedure described by Laemmli (1970). Protein samples were mixed with loading buffer (50 mM Tris, 2% SDS, 2% mercaptoethanol, 15% sucrose, 1% bromophenol blue) and boiled in a water bath for 2 min. The samples were then applied onto gels and subjected to electrophoresis (100 mA for 1 to 1.5 h). Gels were fixed in 1:1 v/v methanol-water mixture and stained with silver. The gel bandings of three replicates and its protein peaks were compared using molecular weight (MW) standards and an Imagery Analysis System (Biocom 500, program Melphor, Les Ulis, France). Only the polypeptide bands present in the three replicates, in the untreated controls as well as in the treated samples, were considered. The MWs were: rabbit muscle phosphorylase b (97.4 kDa), bovine serum albumin (66.2 kDa), hen egg white ovalbumin (42.7 kDA), and soybean trypsin inhibitor (21.5 kDA) (see Fig. 2A).

Results

Polypeptide content. The variations observed in total protein content during the four stages of palynogenesis were similar in the tapetum, in the microspores or in the pollen grains. After the first mitosis, a slight decrease in the protein content was observed in young pollen, then the content increased to reach a maximum in the maturing pollen, and decreased until the mature pollen (Fig. 1). However, the differences in protein content were negligible.

Changes in protein patterns in *Tradescantia* tapetum during pollen development. The protein peaks of the electrophoretic profiles were characterized using molecular weight (MW) standards and the Imagery analysis system. The MW standards are shown in Fig. 2A, described in detail in the chapter Material and Methods.

The peaks of the untreated controls were unevenly distributed through the gel (Fig. 2B). Only two polypeptid bands (16 and 34 kDA) were continuously present in the tapetum of the control plants throughout pollen development (Fig. 3). Another band (38 kDA) was present during three stages but most of the bands were present only at one or two. However, large variations were observed: high MW polypeptides disappeared at stages MP and P, while specific sets of bands appeared (20–30 kDA, MP; 40–50 kDA, P).

Effect of 2,4-D treatments on protein patterns. The number of peaks in treated plants was lower than in the controls and no peak was observed in the high MW range. Some peaks persisted in the region of low MWs while others disappeared

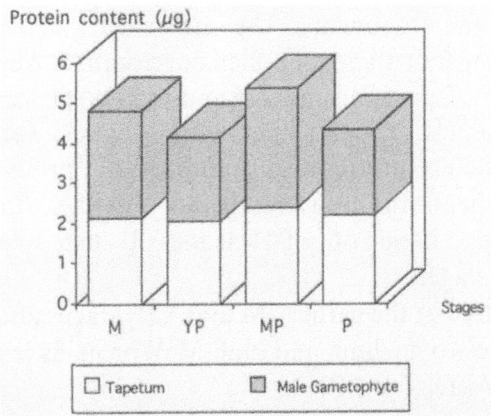

Fig. 1. Protein contents of male gametophytes and corresponding plasmodial tapetum of *Tradescantia bracteata*. *M* microspores, *YP* young pollen, *MP* maturing pollen, *P* mature pollen

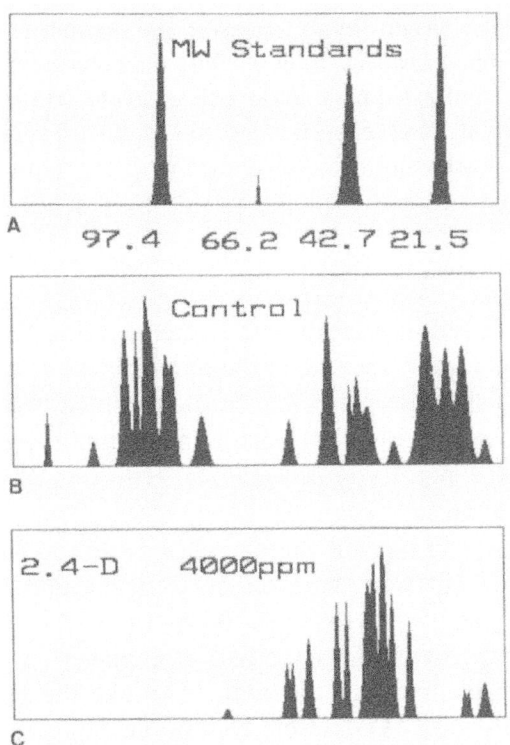

Fig. 2. Peak characterization of the molecular weight standard (*A*), the control (*B*) and treated plants (*C*) in the *Tradescantia bracteata* tapetum at the microspore stage from electrophoretic protein patterns

and new ones appeared (Fig. 2C). The electrophoretic patterns of the tapetum proteins were modified by the 2,4-D treatments applied to *Tradescantia* inflorescences at rates of 4000 and 8000 ppm (Fig. 4). Some polypeptid bands, which were present in the control, were also observed in the treated plants, but in addition new bands appeared.

Polypeptid bands maintained after the treatments: The major part of the proteinic bands present in the control at the four stages of pollen development were also observed in the treated plants. Thus, the 2 protein bands (16 and 34 kDa) present at all stages were not modified by the 2,4-D treatments. Other bands were maintained only by one treatment. The 8000 ppm treatment modified the protein banding less than the 4000 ppm treatment. In both treatments together, the protein pattern was in many cases altered: 3 bands out of 11 at the YP stage were maintained against 5 out of 9 at the MP stage.

The polypeptid bands which disappeared at the earlier [M and YP] stages after the treatments, corresponded in most cases to medium and high MW proteins (e.g. 92,7 and 75 kDa for M; 89,2, 64 and 39,4 kDa for YP).

New polypeptid bands induced by the treatments: Among the new polypeptidic bands observed in the tapetum, one (62 kDa for 8000 ppm 2,4-D) was present during all stages of pollen development, two were common to the first three stages (23 kDa, 4000 ppm; 74 kDa, 8000 ppm) and a 24 kDa band is present at M, YP and P stages

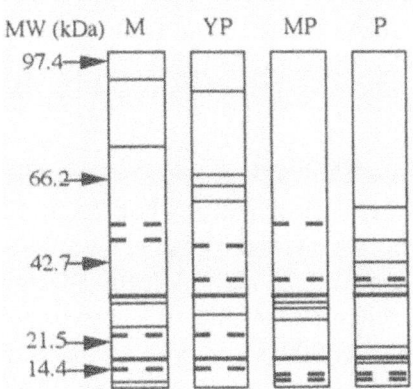

Fig. 3. Protein pattern variation in the *Tradescantia bracteata* tapetum during pollen development in control plants —— Band specific of one stage, – – bands common to two or three stages, —— bands common to all stages

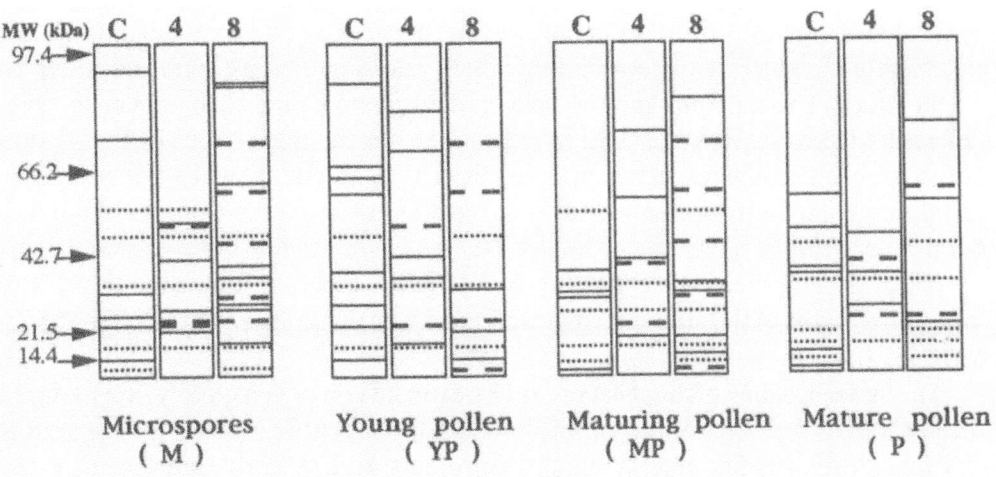

Fig. 4. Modifications of the *Tradescantia bracteata* tapetum protein patterns induced by two doses of 2,4-D. *C* control, *4* 4000 ppm, *8* 8000 ppm. —— Band specific of 1 treatment at 1 stage, – – new bands induced by treatments at several stages, ... bands common to treated and control plants

after 8000 ppm treatment. Most of the new bands observed in treated plants were specific of one stage and one treatment.

Discussion

No significant variation of the total protein content was detected in *Tradescantia* plasmodial tapetum during pollen development. In comparison the protein patterns change because protein synthesis, breakdown or consuming occurs. This result gives evidence for continuous protein synthesis in the tapetum until degeneration. On the contrary, in the tapetum of *Rhoeo discolor*, another *Commelinaceae* close to *Tradescantia*, autoradiography had shown only a late synthesis of proteins (ALBERTINI 1971).

The modifications observed in the electrophoretic protein patterns of the tapetum during pollen development are likely to be related to variations in the activity level of the specific mRNAs which were localized in the tapetum by in situ hybridizations (MASCARENHAS 1990), as their activity is correlated both with the development and the senescence of the tapetum.

The disappearance of high molecular weight protein bands at the onset of pollen maturation affirms the reorientation of the tapetal protein metabolism in the plasmodial tapetum of *Tradescantia*, which had been observed by MEPHAM & LANE (1969) starting from the first mitosis. The disappearance of high molecular weight proteins in the tapetum coincides with a period when rRNA and tRNA syntheses are discontinuous in the pollen grain (MASCARENHAS 1990).

The application of the hormon-like 2,4-D herbicide (Desormone) induced the appearance of new protein bands, in the tapetum protein patterns and also the disappearance of other bands normally present in the controls (the effect depends on the dose of 2,4D). The chemical stress effects induced by the herbicide in the anther tapetum resulted in a modification of the protein patterns which occurred also after a heat shock in somatic cells cultured in vitro (BARNETT & al. 1980). Besides, XIAO & MASCARENHAS (1985) only observed the disappearance of protein bands in the electrophoretic patterns of germinating pollen grains of *Tradescantia* submitted to a heat shock. The tapetum and the male gametophyte would thus appear to react differently to stress. However in *Tradescantia*, from the microspore to the mature pollen grain stage, the respective protein patterns (data not shown) are modified by 2,4-D treatments in the same way as that reported above for the tapetum (unpubl. data).

Natural or induced male sterility is related to tapetum abnormalities. Early initial alterations of the tapetum will result in lesions of the sporogenous cells (MIAN & al. 1974, MASCARENHAS 1990).

The understanding of the effect of 2,4-D treatments on pollen quality first required the analysis of the effect of the applications on the tapetum. Further investigations will include ^{35}S-methionine labelling of proteins and in vitro translation of the mRNAs extracted from the tapetum in order to characterize the new proteins synthesized after the treatments.

The investigations received financial support from ASEDIS – SO (Association Inter-professionnelle du Développement de Semences du Sud Ouest; an association of seed companies) Toulouse, France.

References

ALBERTINI, L., 1971: Les acides nucléiques et les protéines au cours de la microsporogénèse chez *Rhoeo discolor* (HANCE). Etude autoradiographique et cytophotométrique. – Rev. Cytol. Biol. Vég. **34**: 49–92.

– SOUVRÉ, A., AUDRAN, J. C., 1987: Le tapis de l'anthère et ses relations avec les microsporocytes et les grains de pollen. – Rev. Cytol. Biol. Vég. Bot. **10**: 211–242.

AL NAJJAR, N. R., SOLIMAN, A. S., 1982: Cytological effects of herbicides. – Cytologia **47**: 53–61.

BARNETT, T., ALTSCHULER, M., MCDANIEL, C. N., MASCARENHAS, J. P., 1980: Heat shock induced protein in plant cells. – Develop. Genetics **1**: 331–340.

BEDINGER, P., 1992: The remarkable biology of pollen. – Pl. Cell **4**: 879–887.

HUME, L., SHIRRIF, S., 1989: The effect of 2,4-D on growth and germination of lamb's quaters. – Canad. J. Pl. Sci. **69**: 897–902.

LAEMMLI, U. K., 1970: Cleavage of structural protein during the assembly of the head of bacteriophage T4. – Nature **227**: 680–685.

MARTIN, D. A., MILLER, S. D., ALLEY, H. P., 1989: Winter wheat (*Triticum aestivum*) response to herbicides. – Weed Technol. **3**: 90–94.

MASCARENHAS, J. P., 1990: Gene activity during pollen development. – Ann. Rev. Pl. Physiol. Pl. Mol. Biol. **41**: 317–338.

MEPHAM, R. H., LANE, G. R., 1969: Formation and development of the tapetum periplasmodium in *Tradescantia bracteata*. – Protoplasma **68**: 175–191.

MIAN, H. R., KUSPIRA, J., WALKER, G. W. R., MUNTJEWERFF, N., 1974: Histological and cytochemical studies on five genetic male sterile lines of barley *Hordeum vulgare*. – Canad. J. Genet. Cytol. **16**: 355–379.

MULCAHY, D. L., 1986: In recognition of the forgotten generation. – In MULCAHY, D. L., MULCAHY, G. B., OTTAVIANO, E., (Eds): Biotechnology and ecology of pollen, pp. vii–xi. – New York, Berlin, Heidelberg, Tokyo: Springer.

PACINI, E., FRANCHI, G. G., HESSE, M., 1985: The tapetum: its forms, function, and possible phylogeny in *Embryophyta*. – Pl. Syst. Evol. **149**: 155–185.

PENNER, D., ASHTON, F. L., 1966: Biochemical and metabolic changes in plants induced by chlorophenoxy herbicides. – Res. Rev. **14**: 39–113.

TIWARI, S. C., GUNNING, B. E. S., 1986: An ultrastructural, cytochemical and immuno-fluorescence study of postmeiotic development of plasmodial tapetum in *Tradescantia virginiana* L. and its relevance to the pathway of sporopollenin secretion. – Protoplasma **133**: 100–114.

VAUGHN, K. C., LEHNEN, L. P., 1991: Mitotic disrupter herbicides. – Weed Sci. **39**: 450–457.

XIAO, C. M., MASCARENHAS, J. P., 1985: High temperature induced thermotolerance in pollen tubes of *Tradescantia* and heat shock proteins. – Pl. Physiol. **78**: 887–890.

Address of the authors: V. RASOLONJATOVO and A. SOUVRÉ, Laboratoire de Biotechnologie et Amélioration des Plantes (EA-DRED 832), C. BALAGUE, Laboratoire des Industries Agro-Alimentaires, INP-ENSAT, 145 Av. de Muret, F-31076 Toulouse Cédex, France.

Pl. Syst. Evol. [Suppl.] 7: 107–116 (1993)

Calcium and calmodulin distribution in the tapetum of *Gasteria verrucosa* during anther development

M. T. M. WILLEMSE

Key words: *Liliaceae, Gasteria.* – Membrane-bound calcium, calmodulin, tapetum.

Abstract: Calcium and calmodulin are present in the tapetal cells during anther development in *Gasteria*. The highest levels of calcium and calmodulin are shown during the microspore stage just when the microspores show a decrease in calcium and calmodulin content. The calcium and calmodulin levels in the tapetal cells decrease at young pollen phase coinciding with an increase of calcium and calmodulin in the cytoplasm of young pollen. It is postulated that the tapetum stores and supplies calcium during anther development. Trabeculae formation (wall thickenings in the endothecium) is also marked by the presence of calcium.

Calcium is involved in a high number and variety of cellular processes. Ca^{2+} is linked to genetic regulation (GUILFOYLE 1989, BUSH 1992, KAMINEK 1992), to different types of synthesis (POOVAIAH & VELUTHAMBI 1986, KAUSS 1987), to transport (DELA FUENTE 1984, PENEL & al. 1986, STEER 1988), and to development (MARMÉ & DIETER 1983, POOVAIAH 1985, MUTO & HIROSAWA 1987, HEPLER 1989, TIMMERS 1990). This variety of functions makes it difficult to state a special role in the cell or tissue.

Using fluorochromes, such as chlorotetracycline reacting on free calcium present in the vicinity of hydrophobic sites as membranes (CTCf) of fluphenazine which reacts with the calcium/calmodulin complex, designated as activated calmodulin (FPZf), cellular bound calcium can be demonstrated (TSIEN 1989, TIRLAPUR & WILLEMSE 1992). High intensities are related to high quantities and this may be related to a special function of the calcium present.

During pollen germination Ca^{2+} gradients as shown by CTCf are present in the tip of the pollen tube and these are thought to be related to cell tip growth (REISS & HERTH 1978, POLITO 1983, MILLER & al. 1982). A similar tip gradient of CTCf present

Abbreviations:
CTCf = chlorotetracyline fluorochrome
FPZf = fluphenazine fluorochrome
CTC = chlorotetracycline
FPZ = fluphenazine
UVCSLM = UV confocal laser scanning microscope

in germinating pollen of *Gasteria* is already prepared from the tetrad phase of the microspore and is maintained during pollen development (TIRLAPUR & WILLEMSE 1992). The gradual polar position of the CTCf and FPZf during pollen development points to a function of calcium in cell tip growth.

During microsporogenesis and pollen formation, also the tapetal cells show CTCf fluorescence (TIRLAPUR & WILLEMSE & al. 1992). An ultrastructural study of the tapetum development of *Gasteria verrucosa* by KEIJZER & WILLEMSE (1988a,b) reveals the endomitosis and increase in cytoplasm until the tetrad phase. During the development of the microspores the tapetal cells get a vacuole and the formation of pollenkitt begins in the late microspore phase till the mid pollen phase.

The question arises if the calcium bound in the tapetal cells is also involved in the supply of calcium during microsporogenesis and pollen development. In the present study, using CTCf and FPZf during different phases of microspore and pollen development, the presence of bound calcium in the tapetum was related to the microspore and pollen development. An arbitrary and comparative quantification is added to compare the levels of CTCf and FPZf between developing microspores and pollen versus the tapetal cells.

Material and methods

The plant material of *Gasteria verrucosa* (MILL.) H. DUVAL was collected and cultured as described in TIRLAPUR & WILLEMSE (1992). The male sterile *Gasteria* plant M35 was included. Experimental conditions are the same as in TIRLAPUR & WILLEMSE 1992. To compare the treatments with CTC and FPZ sections were taken from the same bud.

The fluorescence intensity of CTCf and FPZf staining, from each selected phase of at least five different sections, was calculated comparing the fluorescence of the meiotic cell, microspore, and pollen grain with that of the tapetal cells. This CTCf or FPZf intensity of the tapetal cells was calculated as less (0), equal ($+$) or more ($++$) in relation to a meiotic cell, microspore or pollen grain. This arbitrary approach to state the tapetum quantity of fluorescence was related to the cytophotometric CTCf and FPZf quantifications of pollen development by TIRLAPUR & WILLEMSE (1992, see Fig. 3). A cytophotometric measurement of the tapetum is possible but difficult to relate to one tapetal cell only. For confocal scanning laser microscopy the Biorad system MRC 600 was used with an adapted equipment for an UV laser.

Results

Localization of CTCf and FPZf in the tapetum during anther development. In the anther the tapetal cells and microspores or the pollen grains are marked by staining of CTC and PFZ. In the other anther tissues the plastids show a red autofluorescence especially in the outer subepidermal layers. Cell walls can get secondary fluorescence due to the staining procedure (Fig. 1a).

In the sporogenic phase only the sporogenous cells react with CTC or FP. Tapetal cells do not show only differentiation (Fig. 1b,c). During pachytene-diplotene some cells of the tapetum show a CTCf (Fig. 1d). After treatment with FPZ seldom a fluorescent tapetal cell can be recognized. The intensity of the FPf in such cells is weak compared with the CTC treated section (Fig. 1e). During the early tetrad stage nearly all tapetal cells show CTCf. The round or flattened cells are positioned against the middle layer (Fig. 1f). The same phase reacts also after the FP treatment. The FPf is visible in flat tapetal cells (Fig. 1g). Early microspores are surrounded with enlarged highly vacuolated tapetal cells. These cells show a bright CTCf and FPZf (Fig. 1h,i). With both methods, the fluorescence is especially intensive in the cytoplasm which surrounds the non-fluorescencing vacuole. In the mid and late microspore stage, the tapetal cells shrink partially due to the loss of the vacuolar content. The tapetal cell layer becomes small and in the cells the CTCf is inensive (Fig. 1j). The FPZf is also intensive but starts to decrease (Fig. 1k). In the young and early pollen grain phases the CTC treated material shows a very faint fluorescence in the tapetum (Fig. 1l). Some of the endothecium cells (Fig. 1l) and a single epidermal cells can show CTCf (Fig. 1n). The flat tapetal cells still have a weak fluorescence after the FPZ treatment. A few epidermal cells show sometimes FPf in the basal part of the cell (Fig. 1m). In the late bicellular pollen grain stage no fluorescent reactions in the degenerated tapetum can be detected in CTC (Fig. 1o) and FPZ (Fig. 1p) treated anthers. Sometimes a single endothecium or epidermis cell may react on CTC treatment (Fig. 1o). The localization of the FPZf and CTCf in the tapetum is presented schematically and related to the pollen development in Fig. 3.

Localization of CTC and FPZ in sterile anthers. In anthers in which the microspore development stops, the tapetum remains present. During the tetrad stage these tapetum cells still react after the CTC as well as FPZ treatments with a fluorescence pattern that is comparable with normal anthers. During the young microspore stage the enlarged cells are still present and also show an intensive CTCf and FPZf (Fig. 1q,r). In these cells the distribution of fluorescence around the vacuole is less distinct. In the mid microspore stage the tapetum cells are present but the CTC treatment does not give a reaction in this tapetum. The endothelium layer however shows clear CTCf as well as some epidermal cells, a reaction comparable with the normal anther development (Fig. 1s). In the mid microspore stage the FPZf is positive in the tapetum. The FPZf is very granular and seems to be restricted to the plastids (Fig. 1t). This granular appearance is less expressed in the normal condition.

UV-confocal laser scanning microscopic localization of CTCf and FPZf in microspores, pollen grains, and tapetal cells during the microspore phase. In a CTC stained microspore the fluorescence is located near the colporal region. Using epifluorescence microscopy few details can be observed. With UV Confocal Laser Scanning Microscopy (UVCSLM) a more detailed localization can be obtained. Figure 2a shows the strong reactions near the colporal zone of a vacuolated microspore in which some plastids show CTCf. The nuclear membrane reacts also with CTC as do the nucleoli. In the cytoplasm a faint localization of the tonoplast and (probably) other membranes is visible. The FPZf is weakly distributed polarly in a young pollen grain. The vegetative nucleus is surrounded by a fluorescent cytoplasm, in which a small vacuole is still present. In the young generative cell the cytoplasm contains some vacuoles, the vegetative cytoplasm shows a positive fluorescent reaction (Fig. 2b).

In the tapetal cells during microspore stage the CTCf visualized with epifluorescence microscopy shows a positive reaction of the cell but only some plastids are visible (Fig. 2c). The FPf of such tapetal cell is more diffuse but can also show

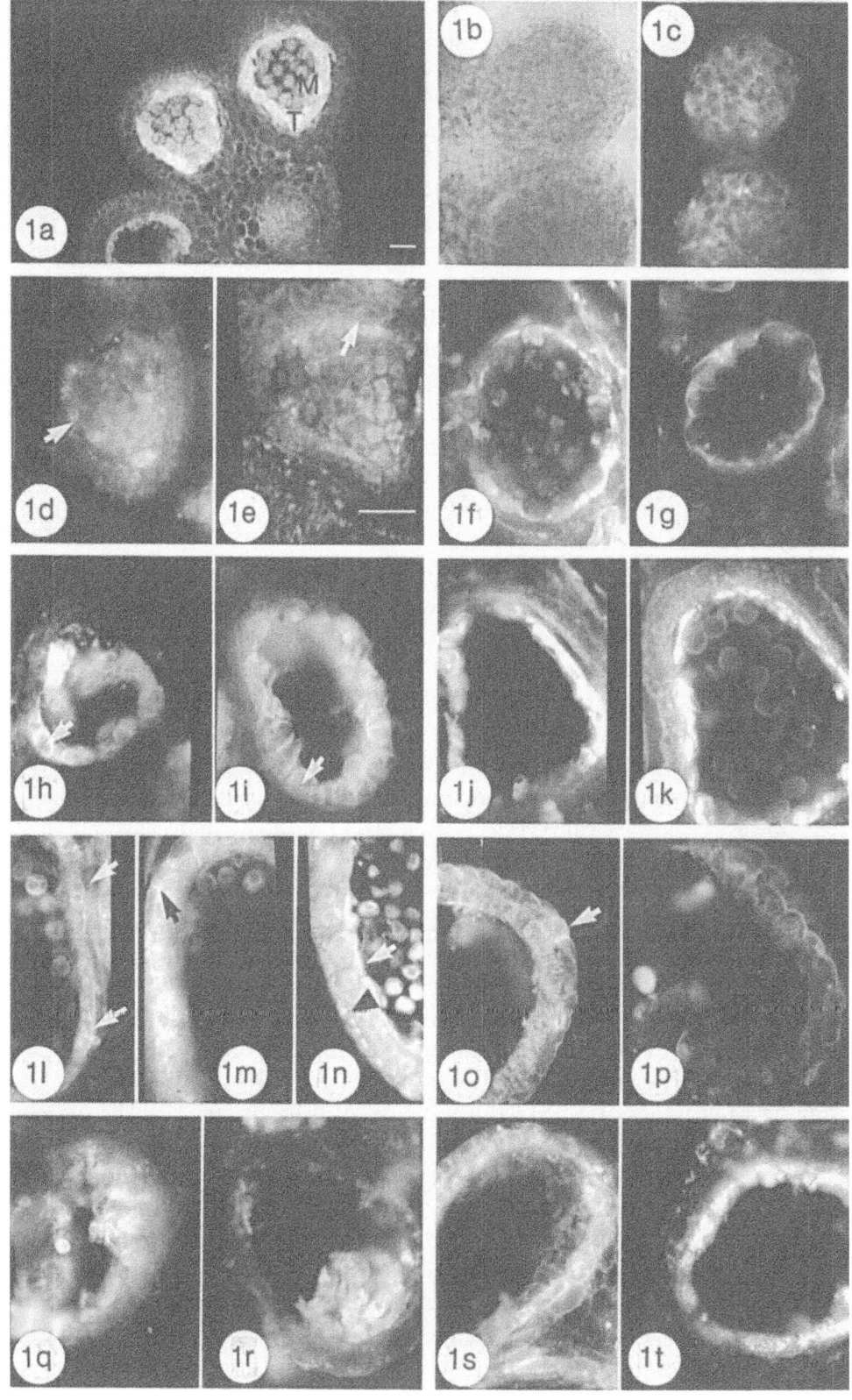

plastids which have an intensive fluorescence (Fig. 2d). With UVCSLM the CTCf is more distinct. Although a cell border is not very distinct, a strong reaction of the plastids is obvious. The nuclear membrane can partially be distinguished and the nucleolus also shows CTCf. In the cytoplasm some faint markings are visible. The vacuole shows no reaction (Fig. 2e). The FPZf is dispersed and only an accumulation in the plastids is evident (Fig. 2f).

Quantitative approach of the CTCf and FPZf in the tapetum. Using atomic absorption and fluorescence intensity measurements during different phases of pollen development TIRLAPUR & WILLEMSE (1992) showed an increase in calcium content in the tetrad stage followed by a decrease in the microspore stage and again an increase in the bicellular pollen grain stage. Peaks of high calcium contents implicate also an influx of calcium in the anther that occurs twice. Between these two periods the tapetum cells enlarge and show a high CTCf of FPZf as described before (see Fig. 3).

In the tapetum the CTCf intensity increases from meiosis until the mid microspore phase and thereafter decreases quickly. The endothelium and some cells of the epidermis show a weak CTCf from early pollen stage on. In the tapetum the

◀ —————————————————————————————————————

Fig. 1. *Gasteria verrucosa. a* Anther cross section during the tetrad phase, treated with CTC, showing a reaction on the tapetum (T) and microspores (M). Anther wall plastid autofluorescence is red and the cell walls have a yellow secondary fluorescence. These red and yellow signals are present on most of the figures and give a background which can be discerned by the high intensity of the bright yellow CTCf or FPZf. Bar: 100 μm. *b, c* Cross section of fresh anther material (*b*) with sporogenous cells without staining, tapetal cells are hardly recognizable. The same section (*c*) showing FPZf in the sporogenous cells. FPZf of future tapetal cells cannot be seen. × 100. *d, e* CTC treated anther with pachytene cells (*d*), some tapetal cells show CTCf (arrow). In a FPZ treated section (*e*) some very faint FPZf in the tapetal cells can be detected (arrow). × 100. Bar for *b–t*: 100 μm. *f, g* In the tetrad phase a band of CTCf (*f*) showing tapetal cells can be recognized. FPZ treatment shows a comparable reaction (*g*) but the tapetal cells fluoresce partly. × 100. *h, i* CTCf (*h*) of the vacuolated and enlarged tapetal cells during early microspore phase. Vacuole: arrowed FPZf (*i*) is bright in the cytoplasm around the vacuole (arrow) of the enlarged tapetal cells. × 100. *j, k* CTCf of the early and late microspore phase (*j*). The tapetal cells are decreased in volume. FPZf of the same phase (*k*) with bright intensity of the contracted cytoplasm. × 100. *l–n* CTC treated early bicellular pollen stage (*l*). The tapetal cells do not show CTCf, only some endothelial cells react (arrows). CTC treated late bicellular pollen stage (*m*) shows some epidermal cells with autofluorescence (arrow). FPZ treated early bicellular pollen grain (*n*). A very weak reaction of the tapetal cell remains (arrow). Also some epidermal cells show a weak reaction (arrowhead). × 100. *o, p* CTC treated mature pollen phase (*o*), the tapetum is degenerated. The anther wall shows a positive reaction in the epidermal cell (arrow). The same mature pollen phase FPZ treated (*p*): The anther wall shows no reaction. × 100. *q, r* Sterile anther treated with CTC (*q*) in the early microspore stage. The tapetal cells show bright fluorescence but a clear cytoplasmic localization around a vacuole cannot be observed as in *h, i*. The same but treated with FPZ (*r*). The cells react positively and comparably with the CTC treated cells. × 100. *s, t* Early microspore phase in a sterile anther. CTC treatment (*s*) gives only a positive signal in the endothecial cells but the tapetum has lost its fluorescence compared to *j*. The same but treated with FPZ (*t*). The tapetum shows a diffuse and a granular fluorescence originating from the plastids. × 100

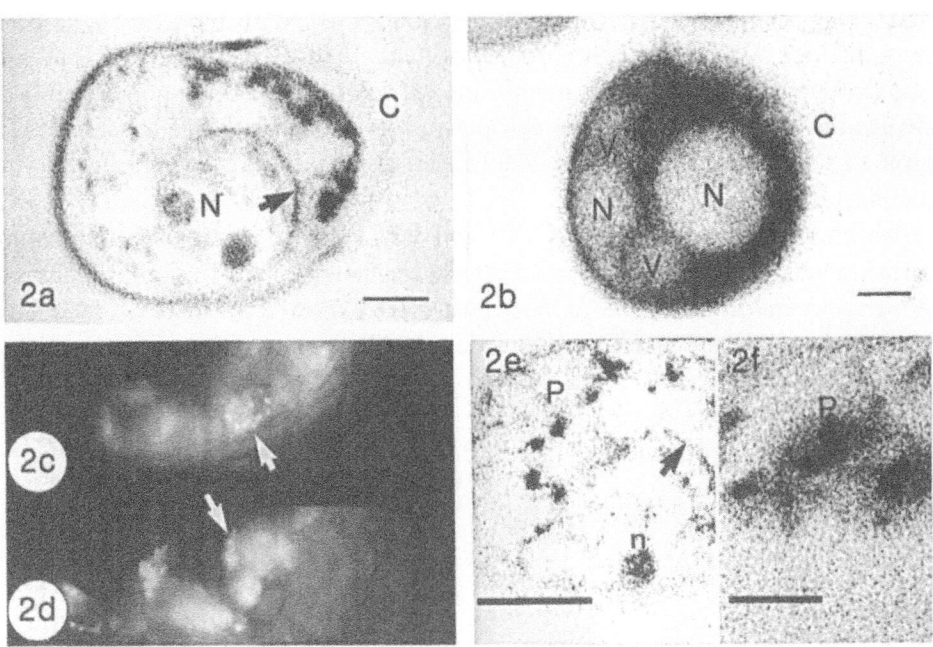

Fig. 2. *Gasteria verrucosa. a* CTCf of a microspore with UVCSLM represented in negative.
Near the colpus (C) a strong reaction is visible. Plastids are present near the colpus (P). The
nuclear membrane (arrow) and nucleoli (N) show CTCf as well as the tonoplast and the
other membranes in the cytoplasm do. The pollen wall (intine) shows autofluorescence. Bar:
10 µm. *b* FPZf in a young bicellular pollen grain represented in negative. The vegetative cell
shows FPZf in the cytoplasm, the nucleus (N) and vacuole (V) are not fluorescent. In the
generative cell the nucleus (N) and vacuoles (V) do not fluorescence, while the cytoplasm
does. Bar: 10 µm. *c, d* CTCf of tapetal cells (*c*) with UVCSLM during the microspore phase
with a reaction in the cytoplasm and in the plastids, visible as dots (see arrow). FPZf (*d*) of
tapetal cells during microspore phase showing a more diffuse reaction and a distinct one in
the plastids (arrow). × 200. *e, f* CTCf of a tapetal cell during the microspore phase (*e*),
observed with UVCSLM and represented in negative. Plastids (P) show intensive
fluorescence, the nuclear membrane (see arrow) and nucleolus (n) are distinctly visible.
Locally in the cytoplasm a more faint reaction is visible. FPZf of the tapetal cell (*f*) during
the microspore phase observed with the UVCSLM and represented in negative. The FPZf
is diffuse and plastids (P) accumulate the FPZ. Bar: *e*: 10 µm, *f*: 5 µm

FPZf is somewhat later visible than the CTCf and increases in meiosis. A slower
decrease occurs during mid pollen phase and the FPZf remains longer in
the tapetal cells. Only some epidermal cells show a very weak fluorescence.

Discussion and conclusions

Using CTCf, which indicates a high level of free calcium in cell compartments in the
vicinity of hydrophobic sites such as membranes, and FPZf, showing activated
calmodulin, a high level of calcium could be demonstrated in the tapetum cells
during anther development. Activated calmodulin appears somewhat later in the
tapetum cells but remains longer in the tapetum than the CTC-calcium. The

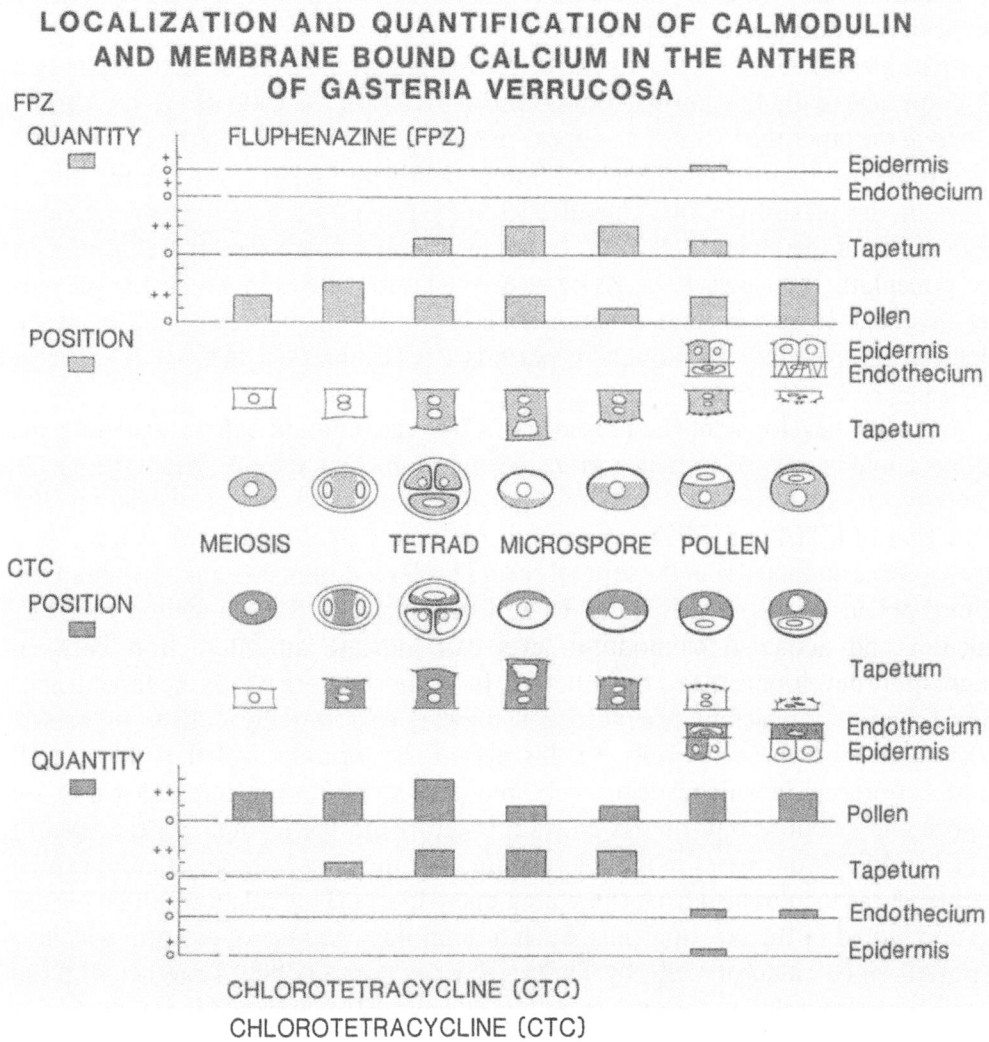

Fig. 3. Schematic representation of the FPZf and CTCf localization in pollen development and tapetal cells of *Gasteria verrucosa*. The histogram represents the estimated fluorescence intensity (o, +, ++) of different cells and tissues

CTC-calcium and calmodulin contents increase simultaneously with the increase of tapetal cell volume. So a gradual increase of CTC-calcium and activated calmodulin from the tetrad stage occurs until the mid microspore stage.

Most of the tapetal functions, e.g., nutrient supply, formation of enzymes (callase), production of sporopollenin precursors, are related to pollen formation and its dispersal (PACINI & al. 1985). In the tapetum calcium might not only be related to the function of the tapetal cells itself but also to the microspore and pollen grain formation. Due to the wide range of cellular processes in which calcium can be involved, it is difficult to point to a special one which characterizes a function of the tapetum. Considering the nutritive character of the tapetum and the export of different products to the loculus, most likely the calcium in the tapetal cell can be involved in these excretion processes.

The CTCf and FPZf as expressed in the sterile anthers indicate a relationship between tapetum and developing microspores. In male sterile *Aloe vera*, a flower comparable with *Gasteria*, the tapetal cell development shows no structural changes as compared with the normal development (KEIJZER & CRESTI 1987). Also in *Gasteria* the tapetum does not disappear immediately. The CTCf in the tapetal cells disappears together with the degeneration of microspores. However, the remaining PFZf in the plastids of the tapetal cells may point to a function of activated calmodulin in the tapetal cell itself. In the microspore phase the tapetal cell forms the pollenkitt (WILLEMSE 1972, KEIJZER & WILLEMSE 1988b) in which the plastids are involved. It may be that calcium is necessary in this process. Such a special function of activated calmodulin in plastids is also observed in *Daucus* embryos (TIMMERS & SCHEL 1992).

In pollen development the presence of CTC calcium and calmodulin near the colpus could be related to pollen grain germination (TIRLAPUR & WILLEMSE 1992). Thereby, the tapetal cells could be involved in exchange and storage of calcium. The reduction of CTCf and FPZf intensity in the young microspore could have as a consequence an increase in the tapetal cells. This level diminishes again when in the young pollen grain the CTCf and FPZf levels increase. Such a switch in CTC-calcium and activated calmodulin level can indicate an interaction between microspore development and the tapetum. In the microspore phase the tapetal cells may function as calcium reservoir. The tapetal cells initially deliver or export calcium to the sporogenic cells. In this period the formation of the callose wall occurs, a process in which calcium is bound (KAUSS 1987). After breakdown of the callose and cellulose wall matrix calcium is set free in the loculus. As the level of CTCf and FPZf decreases in the microspore after the tetrad phase, calcium can be set free in the loculus also from the young microspores. This calcium can be picked up and stored in the tapetum cells. Such a temporary storage of calcium was also reported for oak aleurone cells by TRETYN & KOPCEWICZ (1988). Together with the second calcium influx in the anther both CTCf and FPZf intensity increase. A part of the calcium stored in the tapetum cells can again be transferred to the young pollen and to the endothecium. The CTCf in the endothecium cells coincides with the differentiation of the endothecium wall thickenings (trabeculae). In the small sterile anthers this differentiation continues also. As ROBERTS & HAIGLER (1989) have shown in suspension cultures of *Zinnia*, a change in calcium uptake accompanies the onset of tracheary differentiation. Such relation seems to occur also during the trabeculae formation in the endothelium. The calcium supply to the endothecium in sterile anthers might originate from the second calcium influx in the anther.

The localization of CTCf visualized by UVCSLM results in a distinct pattern in the microspore or tapetal cell. The positive reaction of the nucleolus is remarkable. Although in nucleoli a CTCf signal is present, a hydrophobic condition could be present there during microspore phase. This supports a CTCf gradient and/or polarity of characters as shown for microspore/pollen development by TIRLAPUR & WILLEMSE (1992). The CTCf and FPZf in the tapetal cell plastids is relatively strong, which again points to relevance for pollenkitt formation.

The author wishes to thank Mr S. MASSALT for the photographs, Mr. A. B. HAASDIJK for the scheme and Dr J. H. N. SCHEL and Dr A. C. J. T. TIMMERS for their critical reading of the manuscript.

References

Bush, D. S., 1992: The role of Ca^{2+} in the action of GA in the barley aleurone. – In Karssen, C. M., van Loon, L. C., Vreugdenhill, D., (Eds): Progress in plant growth regulation, pp. 96–104. – Dordrecht, Boston, London: Kluwer.

Dela Fuente, R. K., 1984: Role of calcium in the polar secretion of indoleacetic acid. – Pl. Physiol. **76**: 342–246.

Guilfoyle, T. J., 1989: Secound messengers and gene expression. – In Boss, W. F., Morre, D. J., (Eds): Secound messengers in plant growth and development. Pl. Biol. **6**: 315–326. – New York: Liss.

Hepler, P. K., 1989: Calcium transients during mitosis: observations in flux. – J. Cell Biol. **109**: 2567–2573.

Kaminek, M., 1992: Progression in cytokinin research. – TIBTECH **10**: 159–164.

Kauss, H., 1987: Some aspects of calcium dependent regulation in plant metabolism. – Ann. Rev. Pl. Physiol. **38**: 47–72.

Keijzer, C.J., Cresti, M., 1987: A comparison of anther tissue development in male sterile *Aloe vera* and male sterile *Aloe ciliaris*. – Ann. Bot. **59**: 533–542.

– Willemse, M. T. M., 1988a: Tissue interactions in the developing locule of *Gasteria verrucosa* during microsporogenesis. – Acta Bot. Neerl. **37**: 493–508.

– – 1988b: Tissue interactions in the developing locule of *Gasteria verrucosa* during microgametogenesis. – Acta Bot. Neerl. **37**: 475–492.

Marmé, D., Dieter, P., 1983: Role of Ca^{2+} and calmodulin in plants. – In Cheung, W. Y., (Ed.): Calcium and cell function **4**, pp. 263–311. – New York: Academic Press.

Miller, D. D., Callaham, D. A., Gross, D. J. Hepler, P. K., 1992: Free Ca^{2+} gradient in growing pollen tubes. – J. Cell Sci. **101**: 7–12.

Muto, S., Hirosawa, T., 1987: Inhibition of adventitious root growth in *Tradescantia* by calmodulin antagonists and calcium inhibitors. – Pl. Cell Biol. **28**: 1569–1574.

Pacini, E., Franchi, G. G., Hesse, M., 1985: The tapetum: its form, function, and possible phylogeny in *Embryophyta*. – Pl. Syst. Evol. **149**: 155–185.

Penel, C., Castillo, F. J., Kiefer, S., Greppin, H., 1986: The regulation of plant peroxidases by calcium. – In Trewavas, A. J., (Ed.): Molecular and cellular aspects of calcium in plant development, pp. 365–366. – New York, London: Plenum Press.

Polito, V. S., 1983: Membrane-associated calcium during pollen grain germination: a microfluorometric analysis. – Protoplasma **117**: 226–232.

Poovaiah, B. W., 1985: Role of calcium and calmodulin in plant growth and development. – Hort. Sci. **20**: 347–351.

– Veluthambi, K., 1986: The role of calcium and calmodulin in hormone action of plants: importance of protein phosphorylation. – In Trewavas, A. J., (Ed.): Molecular and cellular aspects of calcium in plant development, pp. 83–90. – New York, London: Plenum Press.

Reiss, H. D., Herth, W., 1978: Visualization of the Ca^{2+} gradient in growing pollen tubes of *Lilium longiflorum* with chlorotetracycline fluorescence. – Protoplasma **97**: 373–377.

Roberts, A. W., Haigler, C. H., 1989: Rise in chlorotetracycline fluorescence accompanies tracheary element differentiation in suspension cultures of *Zinnia*. – Protoplasma **152**: 37–45.

Steer, M. W., 1988: The role of calcium in exocytosis and endocytosis in plant cells. – Physiol. Plant. **72**: 213–220.

Timmers, A. C. J., 1990: Calcium and calmodulin during carrot somatic embryogenesis. – In Sangwan, R. S., Sangwan-Norreel, B. S., (Eds): The impact of biotechnology in agriculture, pp. 215–234. – Dordrecht, Boston, London: Kluwer.

– Schel, J. H. N., 1992: Localization of cytosolic Ca^{2+} during carrot somatic embryogenesis using confocal scanning laser microscopy. – In Karssen, C. M., van Loon, L. C.,

VREUGDENHILL, D., (Eds): Progress in plant growth regulation, pp. 347–353. – Dordrecht, Boston, London: Kluwer.

TIRLAPUR, U. K., WILLEMSE, M. T. M., 1992: Changes in calcium and calmodulin levels during microsporogenesis, pollen development and germination in *Gasteria verrucosa* (MILL.) H. DUVAL. – Sex. Pl. Reprod. **5**: 214–223.

TRETYN, A., KOPCEWICZ, J., 1988: Calcium localization in oak aleurone cells using cholorotetracycline and x-ray microanalysis. – Planta **175**: 273–240.

TSIEN, R. Y., 1989: Fluorescent indications of ion concentrations. – Meth. Cell Biol. **30**: 156–291.

WILLEMSE, M. T. M., 1972: Morphological and quantitative changes in the population of cell organelles during microsporogenesis of *Gasteria verrucosa*. – Acta Bot. Neerl. **21**: 17–31.

Address of the author: M. T. M. WILLEMSE, Department of Plant Cytology and Morphology, Agricultural University Wageningen, Arboretumlaan 4, NL-6703 BD Wageningen, The Netherlands.

Pl. Syst. Evol. [Suppl.] 7: 117–125 (1993)

The significance of the anther tapetum
in the biochemistry of pollen pigmentation – an overview

L. Beerhues, M. Rittscher, H. Schöpker, C. Schwerdtfeger, and R. Wiermann

Key words: *Liliaceae*, *Tulipa* cv. 'Apeldoorn'. – Anthers, tapetum, flavonoid biosynthesis, phenylalanine ammonia-lyase, chalcone synthase, immunofluorescence.

Abstract: Flavonoids such as chalcones, flavonol di- and triglycosides as well as anthocyanins are essentially involved in pollen pigmentation. Phenylalanine ammonia-lyase (PAL) and chalcone synthase (CHS) represent two key enzymes in the biosynthetic pathway of flavonoids. In enzymatical studies the distribution of the two enzymes in anthers of *Tulipa* cv. 'Apeldoorn' was analysed. After separation of the loculus material in a pollen and a tapetum fraction the highest enzyme activities of both PAL and CHS were found in the tapetum fraction whereas the pollen fraction showed only low enzyme activities. This was confirmed by immunohistochemical studies using antibodies against CHS. The CHS was present in the tapetum and epidermal cells of anthers at an early and middle postmeiotic developmental stage. The results imply that the tapetum cells play a crucial role in the flavonoid biosynthesis in the loculus of the anthers and consequently in the pollen pigmentation.

Both flavonoids and carotenoids play a key role in the pigmentation of pollen. The accumulation of flavonoid and/or carotenoid compounds is assumed to serve as an optical attraction of pollinators and as a protection against UV-light (Stanley & Linskens 1974). Very recently it was shown that the flavonoid biosynthesis in anthers is required for development of fertile pollen grains (van der Meer & al. 1992). The anther tapetum is substantially involved in the nourishment of the sporogeneous cells and microspores. Many of male sterile plants possess a non-functional tapetum. This clearly shows the significance of the tapetum cells in micro-sporogenesis.

In the past years it was shown by enzymatical and immunochemical experiments (Herdt & al. 1978, Beerhues & al. 1989) that the tapetum plays a crucial role in the synthesis of phenylpropanes and flavonoids, which are accumulated in or on the structures of the exine (Wiermann & Vieth 1983, Zerback & al. 1989). The results of these experiments are summarized in this overview.

Material and methods

Plant material. The experiments were mainly carried out with anthers of *Tulipa* cv. 'Apeldoorn' and other taxa (see Table 1) cultivated in the Botanical Garden, Münster. In situ localization of CHS was carried out with anthers of *Tulipa* cv. 'Apeldoorn'. Three developmental stages were analysed; stage I: early free microspore stage after complete degradation of the callose; stage II: middle postmeiotic stage, mononucleate pollen grains, the loculus material is intensively yellow pigmented; stage III: late binucleate pollen grains, before anther dehiscence.

Enzyme preparations. To measure PAL and CHS activities the contents (loculus material) of 60–120 anthers were squeezed out into 1 ml of the following buffer: 0.1 M potassium phosphate, pH 6.8, containing 10 mM ascorbic acid, 0.5 mM glutathione, 5 mM thiourea and 0.25 M sucrose. The resulting suspension was separated into a pollen and a tapetum fraction according to HERDT & al. (1978) and BEERHUES & al. (1989).

The isolation and enrichment of tapetum tissue from *Tulipa* anthers and the subsequent determination of PAL activity were performed as described by RITTSCHER & WIERMANN (1983).

Enzyme assays. PAL and CHS activities were assayed by the method of AMRHEIN & ZENK (1971) and BEERHUES & WIERMANN (1988), respectively.

Protein determination. The protein content was determined either according to LOWRY & al. (1951) or BRADFORD (1976).

Antibodies and indirect immunofluorescence. The antibodies used had been raised against CHS from spinach leaves (anti-AI IgG, anti-AII IgG; BEERHUES & WIERMANN, 1988). The in situ localization was done similar to the method of SCHÄCHTELE & STEUP (1986). In contrast to this method the antibodies (anti-AII IgG) were used at a dilution of 1:100 in phosphate-buffered saline (PBS). The cross-sections of anthers were washed in PBS containing 0.5% BSA (w/v). To obtain a reduction of the non-specific background fluorescence in additional experiments the cross-sections were treated with a 1:60 dilution of Eriochrome Black (SIGMA; Deisenhofen, Germany) in PBS (FEY 1972). The preparation was carried out according to SCHÄCHTELE & STEUP (1986); after staining with the FITC-antibodies the cross-sections were washed twice with PBS before they were incubated for 10 to 15 sec with the Eriochrome Black solution. The sections were analysed by epifluorescence microscopy using a LEITZ DIAPLAN with filter block L 3. For the documentation Fujichrome 400 films were used.

Protein blotting and immunostaining of protein blots. These procedures were done as in BEERHUES & al. (1988).

Results

Enzymatical studies. The pollen of gymno- and angiosperms accumulates diverse phenolic compounds. These are present as soluble components on and/or in the structures of the exine (WIERMANN & VIETH 1983, GUBATZ & al. 1986, ZERBACK & al. 1989). Flavonoids are generally accumulated as glycosides. They are mainly di- and triglycosides of kaempferol, quercetin, and isorhamnetin and various anthocyanins (WIERMANN & GUBATZ, 1992).

In the past years it has been well documented that flavonoids are formed from cinnamic acid and acetate units via a chalcone as the first C_{15}-intermediate. The two key enzymes of this pathway are L-phenylalanine ammonia-lyase converting phenylalanine to trans-cinnamic acid and chalcone synthase catalyzing the condensation of one molecule of 4-coumaroyl-CoA and three acetate units from malonyl-CoA to yield naringenin chalcone (2′, 4, 4′, 6′-tetrahydroxychalcone) (Fig. 1). Chalcones are central intermediates in the biosynthetic pathways of the various flavonoid classes (HELLER & FORKMANN 1988). Therefore, PAL and CHS

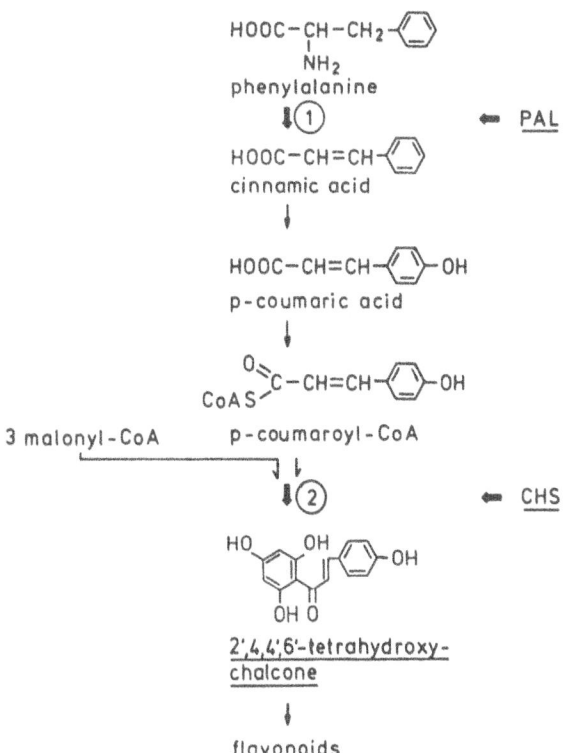

Fig. 1. Simplified scheme of the biosynthetic pathway of flavonoids

represent important markers for the general phenylpropanoid metabolism and the flavonoid pathway, respectively. Their localization yields insight into the compartmentation of flavonoid metabolism in anthers.

PAL and CHS activities of the pollen and tapetum fractions were compared (HERDT & al. 1978, BEERHUES & al. 1989). The tapetum fraction from anthers of *Tulipa* cv. 'Apeldoorn' showed high specific activities of both PAL and CHS, whereas the pollen fraction only contained very low enzyme activities (Fig. 2). Additional experiments showed that the significant distribution of the two key enzyme activities is not restricted to *Tulipa* anthers (Table 1). In all species examined

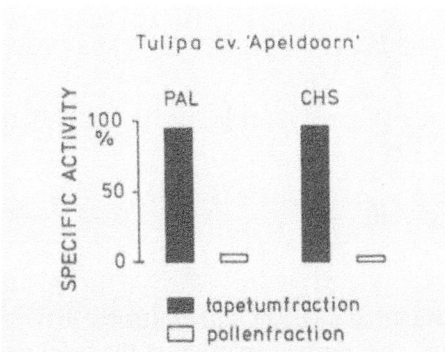

Fig. 2. Distribution of PAL and CHS activities in the pollen and the tapetum fraction of *Tulipa* cv. 'Apeldoorn', obtained by separation of the anther content (after HERDT & al. 1978)

Table 1. Distribution (%) of PAL and CHS activities after pollen/tapetum-fractionation of anther contents from different plant species (after BEERHUES & al. 1989).

	PAL		CHS	
	Tapetum	Pollen	Tapetum	Pollen
Tulipa cv. 'Lustige Witwe'	98.5	1.5	99.2	0.8
Iris pseudacorus L.	96.7	3.3	99.0	1.0
Lilium croceum cv. 'umbellatum'	98.8	1.2	99.1	0.9
Hemerocallis fulva L.	90.5	9.5	100.0	0.0
Narcissus pseudonarcissus L.	95.3	4.7	90.6	9.4

both enzyme activities were located primarily or even exclusively in the tapetum fraction.

Enzymes found in the tapetum fraction obtained by the above mentioned method can be located in the tapetum cells themselves and/or as tapetum derived products in the nutrient solution of the loculus, which surrounds the pollen grains, as well as in the outer structures of the sporoderm. To study the actual localization of one of the two key enzymes, PAL, within anthers of *Tulipa* cv. 'Apeldoorn' (during postmeiotic stage), the specific PAL-activities of different anther tissue preparations were determined (Fig. 3): whole anthers (A), anther cross-sections (0.3 mm) without pollen and the nutrient solution of the loculus (B), the remainder of the anther after squeezing out the loculus material (C), the loculus material (D), pollen (E), and isolated and purified tapetum tissue (F). The starch content was used to characterize the purity of the tissue preparations. The highest PAL activity was found in the tapetum tissue, whereas the pollen extract only showed low or no measurable PAL activity (Fig. 3). The small amount of starch found in the pollen and the

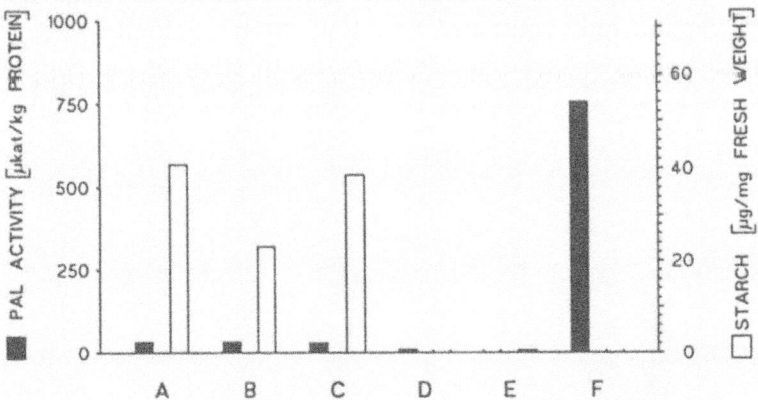

Fig. 3. Distribution of PAL activity and starch content in different anther tissues. *A* Whole anthers, *B* anther cross-sections without pollen and the nutrient solution of the loculus, *C* the remainder of the anther after squeezing out the loculus material, *D* the loculus material, *E* pollen, *F* isolated and purified tapetum tissue (after RITTSCHER & WIERMANN 1983)

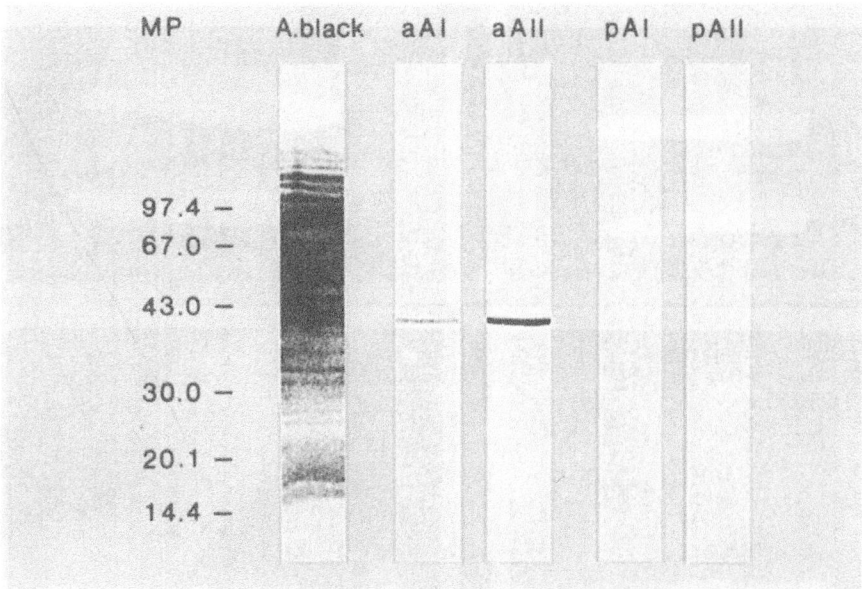

Fig. 4. Immunoblotting of anther content. The SDS-PAGE was performed in a 9–16.5% gradient gel with 180 µg protein applied per lane. The blots were either stained with amido black or incubated with anti-AI IgG (aAI), anti-AII IgG (aAII), pre-immune-AI IgG (pAI), or pre-immune-AII IgG (pAII). *MP* Marker proteins

tapetum tissue fractions indicates the high purity of these fractions. They are not contaminated by cells or cell fragments of the outer anther wall, which showed a high starch content in the developmental stages investigated (RITTSCHER & WIERMANN 1983).

Immunohistochemical studies. Specific antibodies were used to localize CHS in anthers of *Tulipa* cv. 'Apeldoorn' by an immunohistochemical approach. The antibodies had been raised against two chalcone synthase isoforms from spinach leaves. The two antisera [anti-AI IgG (a AI); anti-AII IgG (a AII)] showed a close immunochemical relationship (BEERHUES & WIERMANN 1988). The cross-reaction with CHS protein from tulip anthers was demonstrated by immunoblotting of crude loculus extract after SDS-PAGE. Both antisera detected a single protein band (Fig. 4). The subunits of CHS from *Tulipa* anthers had a molecular mass of about 41 k Da. When a blot was incubated with pre-immune serum no band was specifically stained. Thus, the antibodies cross-reacted monospecifically and were suitable for in situ-localization experiments.

Immunofluorescence studies using aAII were carried out with anthers at an early, middle, and late postmeiotic developmental stage. The tapetum at an early stage of development (stage I) showed intense immunofluorescence (Fig. 5b); tapetum cells of the middle postmeiotic stage (stage II) even stronger immuno-fluorescence (Fig. 5e). Furthermore, immunofluorescence was observed in the epidermal cells, whereas the fluorescence of the residual anther tissue was equivalent to the background level. Sections treated with pre-immune serum exhibited, if at all, only a low uniform background fluorescence (Fig. 5c, f).

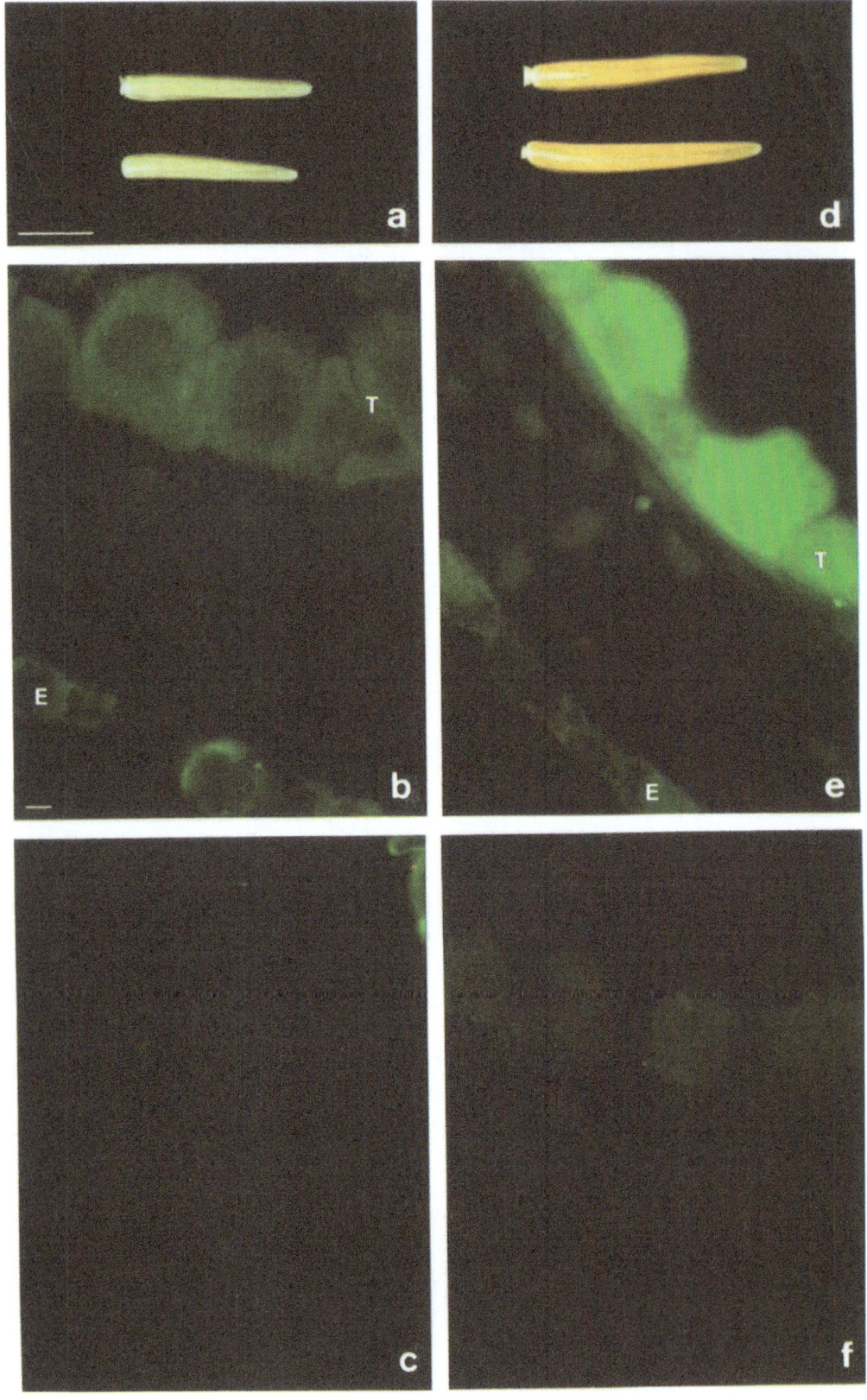

With increasing anther development there was an intense autofluorescence of the tapetum and the pollen walls. This could be largely reduced by treatment with Eriochrome Black. However, at stage III autofluorescence was too strong to be suppressed. Therefore, an unequivocal immunohistochemical proof of CHS was no longer possible.

Discussion

In the course of pollen development in *Tulipa* cv. 'Apeldoorn' a highly differentiated, phase specific accumulation of various soluble phenylpropanoids is observed. In the first phase of the accumulation sequence, after degradation of the callose wall, mono-, di-, and triferuloyl sucrose is formed besides p-coumaric acid conjugates (BÄUMKER & al. 1988). In the second phase, during the middle postmeiotic development differently substituted chalcones are accumulated. The main compound is the 2′, 3, 4, 4′, 6′ – pentahydroxychalcone. In the end phase of pollen ripening di- and triglycosides of kaempferol, quercetin, isorhamnetin, and delphinidin are formed (STRACK & al. 1981) and accumulated as soluble components in and/or on the structures of the outer pollen wall (WIERMANN & VIETH 1983, ZERBACK & al. 1989).

The phase-specific accumulation of different phenylpropanoids is correlated to a considerable extent with the development of enzymes involved in the formation of the individual compounds (WIERMANN 1981). Enzymatic studies with loculus material showed that PAL exhibits high enzyme activity in the early stages of microsporogenesis. CHS shows highest enzyme activity in the middle postmeiotic developmental stages, i.e. stages where intensive chalcone accumulation occurs. The somewhat stronger immunofluorescence observed in stage II upon antibody treatment is therefore in good agreement with the enzyme activity.

Using different methods the key enzymes of the biosynthetic pathway of flavonoids, PAL and CHS, were localized in the tapetum cells of the anther. Thus the tapetum plays a crucial role in flavonoid metabolism in the loculus of the anther and consequently in pollen pigmentation.

These data are in agreement with results of genetical and molecular biological studies. COE & al. (1981) showed that the pigmentation of maize pollen is determined by the genotype of the sporophyte from which pollen originates (compare STYLES & CESKA 1981). Analyses at the molecular level have shown that the expression of genes encoding CHS is restricted to the tapetal cells (KOES & al. 1990).

In the case of CHS immunofluorescence was also seen in the epidermal cells of the anthers. This corresponds well with data described by HRAZDINA & al. (1982),

---

Fig. 5. Anthers of *Tulipa* cv. 'Apeldoorn' at an early (a) and middle (d) postmeiotic developmental stage. Bar: 5 mm. Immunofluorescence localization of CHS in cross sections of tulip anthers (b, c; e, f). Bar: 10 µm. b, c: Cross section of an anther at an early postmeiotic developmental stage incubated with anti AII IgG (b) or pre-immune AII IgG (c): reduction of background fluorescence by a treatment with Eriochrome Black. e, f: Cross section of an anther at a middle postmeiotic developmental stage treated with anti AII IgG (e) or pre-immune AII IgG (f). T Tapetum, E epidermis

Beerhues & al. (1988), Jahnen & Hahlbrock (1988) and Schmelzer & al. (1988), which showed that the leaf epidermis is an important centre of flavonoid biosynthesis.

An intact flavonoid biosynthesis in the anther loculus is an important prerequisite for development of fertile pollen grains. This has been shown by experiments in which the flavonoid biosynthesis was inhibited by an antisense approach using transgenic *Petunia* plants. Anthers of those transgenic plants produced only white pollen without flavonoids, the pollen grains were sterile. Furthermore, the white pollen grains of transgenic anthers were unable to germinate in vitro (van der Meer & al. 1992). The results of these experiments open a new and interesting view on the function of flavonoids.

This work was supported by a grant of the Deutsche Forschungsgemeinschaft and the Fonds der Chemischen Industrie.

References

Amrhein, N., Zenk, M. H., 1971: Untersuchungen zur Rolle der Phenylalanin-Ammonium-Lyase (PAL) bei der Regulation der Flavonoidsynthese im Buchweizen (*Fagopyrum esculentum* Moench). – Z. Pflanzenphysiol. **64**: 145–168.

Bäumker, P. A., Arendt, S., Wiermann, R., 1988: Metabolism of ferulic acid sucrose esters in anthers of *Tulipa* cv. Apeldoorn: I. The accumulation of esters and free sugars. – Z. Naturforsch. **43**c: 641–646.

Beerhues, L., Wiermann, R., 1988: Chalcone synthases from spinach (*Spinacia oleracea* L.). I. Purification, peptide patterns, and immunological properties of different forms. – Planta **173**: 532–543.

– Robenek, H., Wiermann, R., 1988: Chalcone synthases from spinach (*Spinacia oleracea* L.). II. Immunofluorescence and immunogold localization. – Planta **173**: 544–553.

– Forkmann, G., Schöpker, H., Stotz, G., Wiermann, R., 1989: Flavanone 3-hydroxylase and dihydroflavonol oxygenase activities in anthers of *Tulipa*. The significance of the tapetum fraction in flavonoid metabolism. – J. Pl. Physiol. **133**: 743–746.

Bradford, M. M., 1976: A rapid and sensitive method for quantitation of microgram quantities of protein utilizing the principle of protein dye binding. – Anal. Biochem. **72**: 248–254.

Coe, E. H., McCormick, S. M., Modena, S. A., 1981: White pollen in maize. – J. Heredity **72**: 318–320.

Fey, H., 1972: Eriochrome Black, a means for reduction of nonspecificity in immunofluorescence. – Path. Microbiol. **38**: 271–277.

Gubatz, S., Herminghaus, S., Meurer, B., Strack, D., Wiermann, R., 1986: The location of hydroxycinnamic acid amides in the exine of *Corylus* pollen. – Pollen & Spores **28**: 347–354.

Heller, W., Forkmann, G., 1988: Biosynthesis. – In Harborne, J. B., (Ed.): The flavonoids, pp. 399–425. – London: Chapman and Hall.

Herdt, E., Sütfeld, R., Wiermann, R., 1978: The occurrence of enzymes involved in phenylpropanoid metabolism in the tapetum fraction of anthers. – Europ. J. Cell Biol. **17**: 433–441.

Hrazdina, G., Marx, G. A., Hoch, H. C., 1982: Distribution of secondary plant metabolites and their biosynthetic enzymes in pea (*Pisum sativum* L.) leaves. – Pl. Physiol. **70**: 745–748.

JAHNEN, W., HAHLBROCK, K., 1988: Differential regulation and tissue-specific distribution of enzymes of phenylpropanoid pathways in developing parsley seedlings. – Planta **173**: 453–458.

KOES, R. E., BLOKLAND, R. VAN, QUATTROCCHIO, F., TUNEN, A. J. VAN, MOL, J. N. M., 1990: Chalcone synthase promoters in *Petunia* are active in pigmented and unpigmented cell types. – Pl. Cell **2**: 379–392.

LOWRY, O. H., ROSEBROUGH, N. J., FARR, A., L., RANDALL, R. J., 1951: Protein measurement with the folin phenol reagent. – J. Biol. Chem. **193**: 265–275.

MEER, I. M. VAN DER, STAM, M. E., TUNEN, A. J. VAN, MOL, J. N. M., STUITJE, A. R., 1992: Antisense inhibition of flavonoid biosynthesis in *Petunia* anthers results in male sterility. – Pl. Cell **4**: 253–262.

RITTSCHER, M., WIERMANN, R., 1983: Occurrence of phenylalanine ammonia-lyase (PAL) in isolated tapetum cells of *Tulipa* anthers. – Protoplasma **118**: 219–224.

SCHÄCHTELE, C., STEUP, M., 1986: α-1,4 glucan phosphorylase forms from leaves of spinach (*Spinacia oleracea* L.). I. In situ localization by indirect immunofluorescence. – Planta **167**: 444–451.

SCHMELZER, E., JAHNEN, W., HAHLBROCK, K., 1988: In situ localization of light-induced chalcone synthase mRNA, chalcone synthase, and flavonoid end products in epidermal cells of parsley leaves. – Proc. Natl. Acad. Sci. USA **85**: 2989–2993.

STANLEY, R. G., LINSKENS, H. F., 1974: Pollen: biology, biochemistry, management. – Berlin, Heidelberg, New York: Springer.

STRACK, D., SACHS, G., WIERMANN, R., 1981: Pollen of *Tulipa* cv. Apeldoorn as an accumulation site of flavonol di- and triglycosides. – Z. Pflanzenphysiol. **103**: 291–296.

STYLES, E. D., CESKA, O., 1981: Genotypes affecting the flavonoid constituents of maize pollen. – Maydica **26**: 141–152.

WIERMANN, R., 1981: Secondary plant products and cell and tissue differentiation. – In STUMPF, P. K., CONN, E. E., (Eds): The biochemistry of plants. 7: Secondary plant products, pp. 85–116. – New York, London, Toronto, Sydney, San Francisco: Academic Press.

– VIETH, K., 1983: Outer pollen wall, an important accumulation site for flavonoids. – Protoplasma **118**: 230–233.

– GUBATZ, S., 1992: Pollen wall and sporopollenin. – Int. Rev. Cytol. **140**: 35–72.

ZERBACK, R., DRESSLER, K., HESS, D., 1989: Flavonoid compounds from pollen and stigma of *Petunia hybrida*: Inducers of the vir region of the *Agrobacterium tumefaciens* Ti plasmid. – Pl. Sci. **62**: 83–91.

Addresses of the authors: L. BEERHUES, Institut für Pharmazeutische Biologie, Universität Bonn, Nussallee 6, D-53115 Bonn, Federal Republic of Germany. – M. RITTSCHER, H. SCHÖPKER, C. SCHWERDTFEGER, R. WIERMANN, Institut für Botanik und Botanischer Garten, Universität Münster, Schlossgarten 3, D-48149 Münster, Federal Republic of Germany.

Pl. Syst. Evol. [Suppl.] 7: 127–145 (1993)

Recent trends in tapetum research. A cytological and methodological review

Michael Hesse and Michael W. Hess

Key words: Spermatophyta, *Bromeliaceae*, *Tillandsia*, *Tiliaceae*, *Tilia*. – Anther tapetum, preparation artifacts, freeze fixation (high pressure freezing), freeze substitution, abnormal cytoplasmic degeneration, male sterility.

Abstract: Cytological, developmental, and functional aspects of tapetum cells have been extensively studied in the past using classical light- and electron microscopy (LM, EM). Recently, some advanced preparation methods and analytical techniques have come of age; they already have been applied with some success for the investigation of pollen and tapetum development. Analytical transmission EM techniques (ESI, EELS) as well as immunocyto-chemistry and fluorescence microscopy have become invaluable tools for the localization of substances. Laser scanning microscopy and video-enhanced microscopy have brought about a revival of LM, the latter method permitting the study of dynamic cellular processes with high spatial and temporal resolution. Hence, so far static notions of cellular ultrastructure based on the interpretation of classical TEM micrographs can now be completed by a more dynamic concept. In TEM, the application of freeze fixation and freeze substitution as well as sophisticated chemical fixation protocols have considerably improved ultrastructure preservation thereby revealing, for instance, that tapetum senescence starts much later than generally assumed. Some patterns of cytoplasmic disorganization which are usually interpreted as tapetum degeneration were found to result from specimen preparation. Besides presenting recent advances in the fields mentioned above, this paper also reviews literature on aspects of primary/secondary tapetum metabolisms and the regulation of normal/abnormal tapetum (and microspore) development.

For several decades the anther tapetum has been object of interest for many botanists. This is reflected by the relatively large number of reviews in this very specific field of research (Carniel 1963, Echlin 1971, Bhandari 1984, Shivanna & Johri 1985, Pacini & al. 1985, Albertini & al. 1987, Chapman 1987, Pacini 1990, Hesse 1991). However, the recent introduction of advanced techniques in specimen preparation and fixation combined with improved analytical methods (ESI, EELS, immunolabelling) has made it necessary to reinterpret "well known" features of tapetum development and structure. Furthermore, video-enhanced light microscopy (LM) allows to study dynamic cellular processes of which transmission electron microscopy (TEM) only gives a static impression. This is especially important for metabolically highly active cells like pollen tubes and tapetum cells.

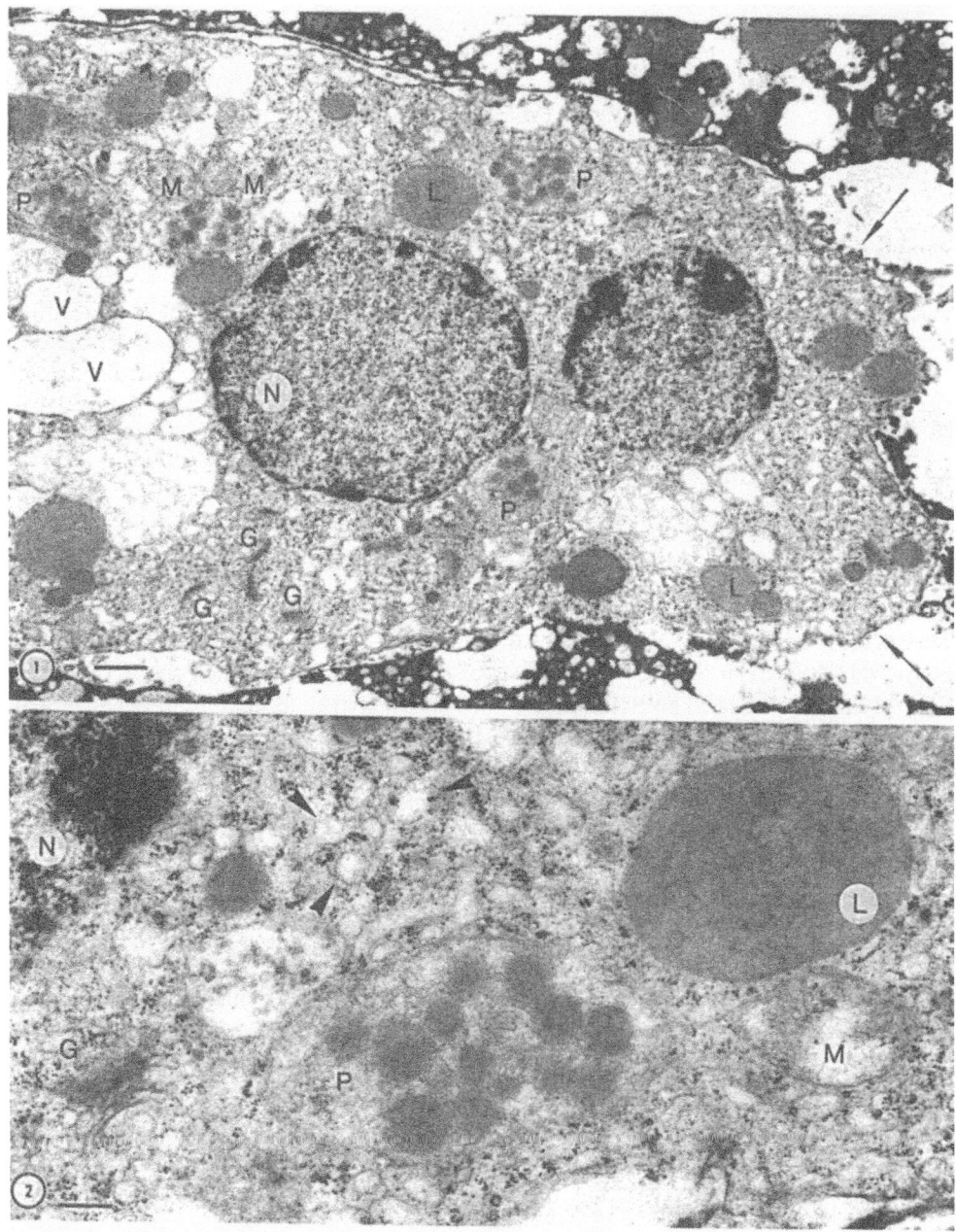

Figs. 1, 2. Anther tapetum of *Tillandsia pallidoflavens* immediately after callose dissolution. Conventional glutaraldehyde/osmium fixation yields inadequate ultrastructure preservation. Nucleus (N), plastids (P), mitochondria (M), endoplasmic reticulum (ER: arrow-heads), Golgi bodies (G), vacuoles (V), plasma membrane (arrows), lipid bodies (L). – Fig. 1: bar = 1 µm, Fig. 2: bar = 0,25 µm

The present paper focuses on two, quite different topics. First, we compare micrographs of freeze-fixed, freeze substituted tapetum cells with chemically fixed ones and discuss methodological problems encountered with (freeze-) fixation of tapetum cells; second, we review recently published papers on anther tapetum metabolism and normal/abnormal tapetum and microspore development.

Material and methods

Plant material. *Tillandsia pallidoflavens* Mez (*Bromeliaceae*) and *Tilia platyphyllos* Scop. (*Tiliaceae*) were grown in the Botanical Garden of Vienna (HBV).

Specimen preparation:

High pressure freezing and freeze substitution. According to Studer & al. (1989) intact anthers of *Tillandsia* were immersed in 1-hexadecene (Fluka, Buchs, Switzerland) and kept for less than 1 minute under mild vacuum to replace intercellular gases (evacuation by hand within a syringe). The use of hexadecene has no visible effects on the viability of various organisms (Studer & al. 1989) which also pertains to microspores and pollen grains (Hess, unpub.). The anthers were subsequently frozen in a Balzers high pressure freezer HPM 010 (Balzers Union, Balzers, Liechtenstein). Freeze substitution in anhydrous acetone containing 1% OsO_4 was performed at about $-90°C$ for at least 8 hours, in a substitution device according to Hess & Glaser (1993). The samples were subsequently allowed to reach room temperature. After 3 rinses in acetone the samples were infiltrated for 3 days with graded resin series (Epon-Araldite) at room temperature and polymerized at 60°C.

Chemical fixation. (a) *Tillandsia*: Anthers were fixed in 2,5% glutaraldehyde (2 hours, 25°C) followed by 1% OsO_4 (2 hours, 4°C). (b) *Tilia*: Anthers were fixed in 3% glutaraldehyde (24 hours, 25°C) followed either by 2% OsO_4 alone (4 hours, 4°C) or by a mixture of 1% OsO_4 and 0,8% $K_3Fe(CN)_6$ (Os-FeCN: 4 hours, 4°C: Hepler 1981 modified by Weber 1991).

Subsequently the anthers of both species were rinsed in distilled water, dehydrated in a graded ethanol series and embedded in Spurr's Low Viscosity Resin.

Electron Microscopy. Ultrathin sections stained with uranyl acetate (40 min/25°C) and lead citrate (5 min/25°C) were examined with a Zeiss EM 109 and a Zeiss EM 900 (Zeiss, Oberkochen, Germany), at 50 kV and 80 kV.

A) Methodological problems resulting from the very special cytological features of tapetum cells

1) The tapetum and other nutritive tissues

The (anther) tapetum is a highly specialized and delicate tissue surrounding the developing spores/pollen grains in most Bryophyta, all Pteridophyta and all Spermatophyta. Pacini & Franchi (1993) attribute 14 different functions to the tapetal cells. Regardless of any morphological or cytological differences tapetum cells can be considered as gland cells synthezising and exuding a lot of diverse enzymes, carbohydrates and secondary substances via various secretory pathways. These metabolic activities do not take place simultaneously. Thus, tapetum cells show several consecutive periods of increasing and decreasing accumulation and/or release of secretory products (Rowley & Walles 1987, Rowley & al. 1992, Fitzgerald & al. 1993b, Hess & Hesse 1993, Rowley 1993, Rowley & Walles 1993). During such particular phases of activity and depending on the nature of the secretory products the cytological features of the anther tapetum closely resemble those of other, highly specialized plant tissues; such are, for instance, secretory cells of the stigma and the stylar transmitting tissue as well as glands producing nectar (cf. Robards & Stark 1988), oils, resins and other secondary substances; further, lipid accumulating tissues in seeds, the nutritive tissues of plant galls as well as the female equivalent to the anther tapetum, the endothelium (or integumentary tapetum) of tenuinucellate ovules. Most of these cell types undergo cytoplasmic degeneration; this process is widely considered to be related to the secretory function of these cells. But what means "degeneration"? Are necrotic patterns observed in fixed cells natural or are they, at least in part, due to specimen preparation? How can we distinguish between preparation artifacts and senescence patterns or tapetum dysfunctions causing male sterility? All these questions are, in our opinion, worth to be discussed in the light of advanced TEM-techniques.

2) Some aspects of ultrastructure preservation

In a recent review ROBARDS (1991) precisely sums up the main problem of conventional TEM when he stated that: "Immobilization and preservation of living cells by chemical fixation is fraught with problems of artifacts." Cellular components may be moved from their *in vivo* position (MINEYUKI & GUNNING 1988, WILSON & al. 1990, KAMINSKYJ & al. 1992), or be completely extracted. Dynamic processes, in particular, are not halted fast enough to guarantee a depiction of their *in vivo* condition with TEM. By contrast, freeze fixation allows immobilization of cellular substructure within milliseconds and has been successfully applied as alternative to chemical fixation (ROBARDS 1991 and references therein). Our three series of micrographs should elucidate the benefits and the limitations of different fixation protocols for preparing tapetum cells for TEM. "Conventional" chemical fixation procedures are compared with high pressure freeze fixation followed by freeze substitution as well as with sophisticated chemical fixation.

Conventional aldehyde/osmium fixation of *Tillandsia* (Figs. 1, 2, 4) and *Tilia* (Fig. 7): The membrane/-bound organelles are present but, in general, not adequately preserved.

Os-FeCN fixation of *Tilia* (Fig. 8): The Os-FeCN fixation protocol yields improved preservation in particular of contents and/or matrix substances of the cell compartments (see also HESSE 1993, this volume).

High pressure freezing, freeze substitution of *Tillandsia* (Figs. 3, 5, 6): All cell types of the anther show excellent ultrastructural preservation irrespective of their developmental stage. The membrane/-bound organelles (plasma membrane, endoplasmic reticulum = ER, Golgi bodies, nuclear envelope, mitochondria, plastids) possess smooth outlines and are usually clearly distinguishable (Figs. 3, 5, 6); only the ER does not always exhibit contrast against the ground cytoplasm (Fig. 3). The

--▶

Fig. 3. *Tillandsia pallidoflavens* subjected to high pressure freeze-fixation and freeze substitution (same stage as Figs. 1, 2). The plasma membrane (arrows) and endomembranes (G, V) appear smooth, plastids (P), mitochondria (M) and the nucleus (N) are turgescent. The matrix of plastids and mitochondria as well as the contents of the cisternae and/or secretory vesicles of the Golgi bodies (G) and vacuoles (V) are intensively stained. The locular fluid (asterisks) and cytoplasmic pollenkitt precursor substances (stars) are also well preserved. The ER, however, is hardly visible at this developmental stage, a common feature in freeze substituted plant cells (cf. ROBARDS & STARK 1988). The plastids contain electron lucent lipophilic inclusions and starch granules which are not easily distinguishable from each other after uranyl acetate/lead staining. – Bar = 0,5 µm

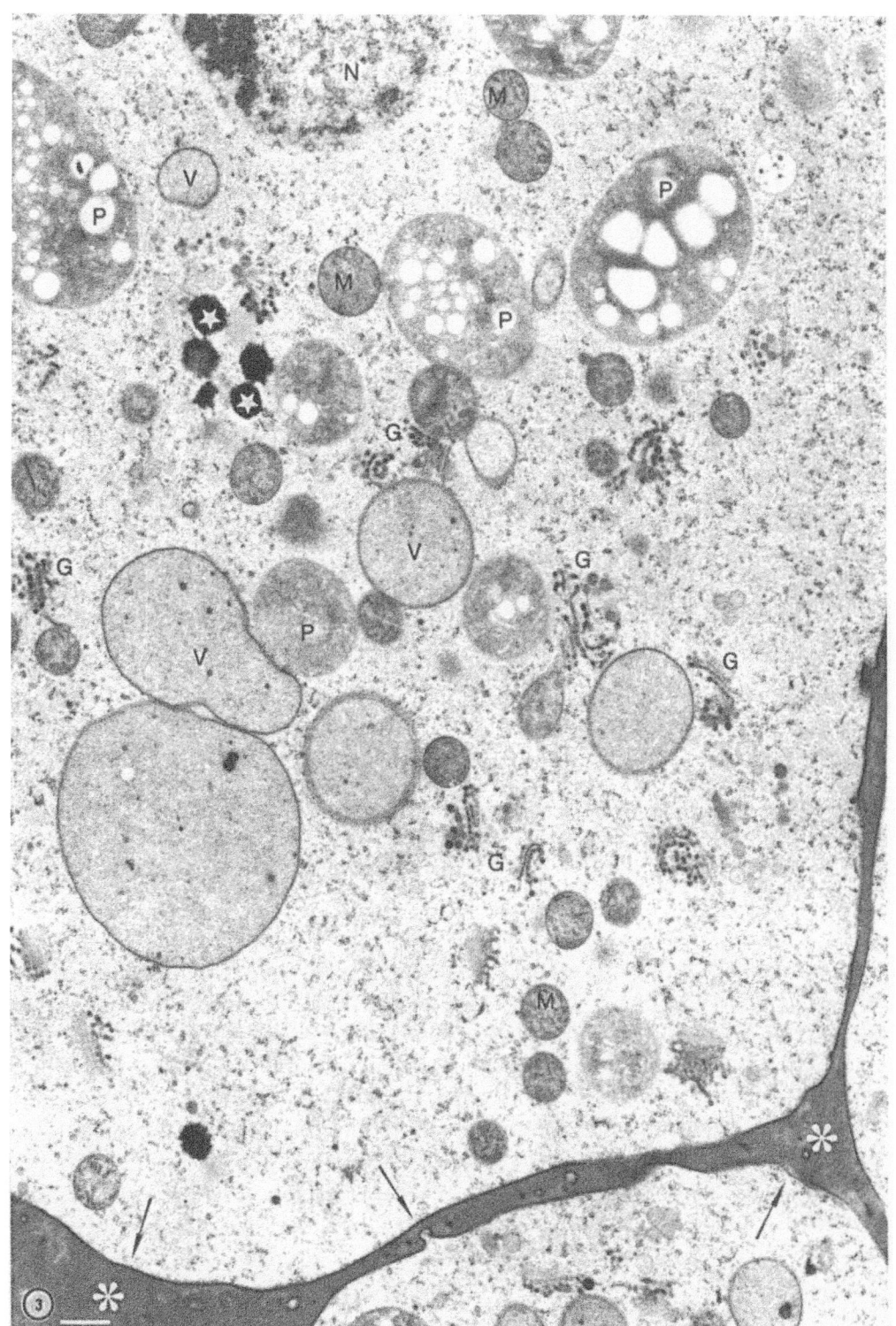

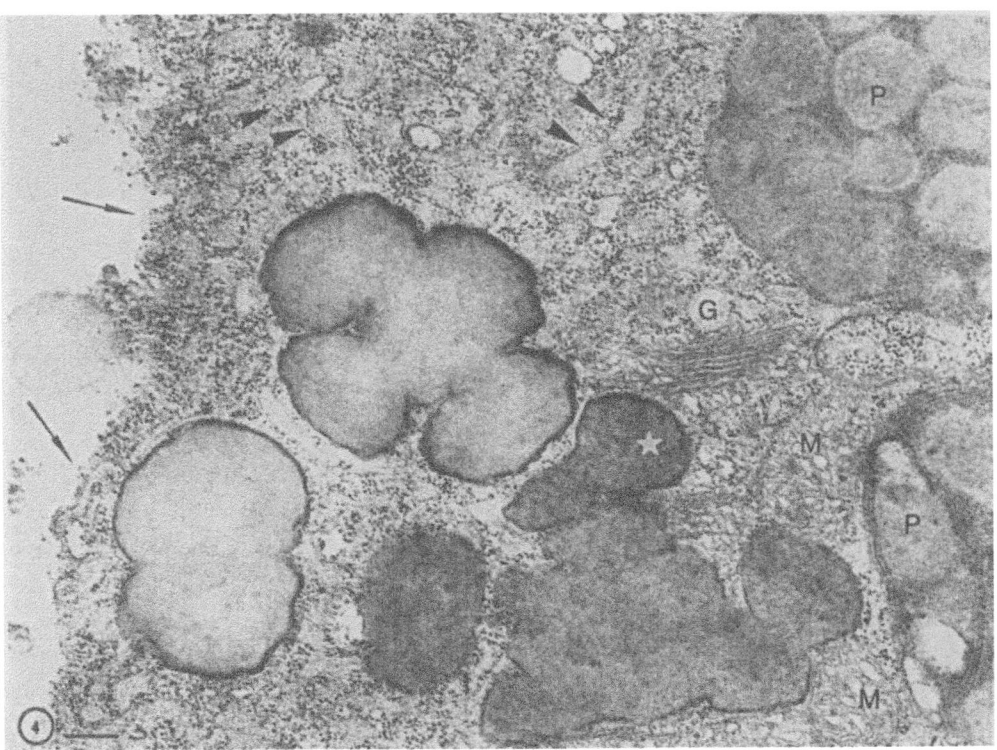

Fig. 4. Anther tapetum of *Tillandsia pallidoflavens* prior to microspore mitosis. Conventional glutaraldehyde/osmium fixation. Large lipid bodies (L) and plastids (P) containing pollenkitt precursors are the only structural details to be unequivocally identified. Mitochondria (M), ER (arrow-heads) and Golgi bodies (G) are only occasionally discernable. The plasma membrane (arrows) appears undulated due to shrinkage during specimen preparation. – Bar = 0,25 μm

matrix of plastids and mitochondria, the contents of vacuoles and secretory vesicles are well preserved (Figs. 3, 5, 6). Microtubules frequently occur throughout the tapetum cells (not shown), whereas microfilaments have not been observed.

The poor preservation quality of glutaraldehyde/osmium fixed tapetum cells at the free microspore stage (Figs. 1, 2, 4, 7) could easily be misinterpreted as cytoplasmic disorganization, i. e. tapetum senescence, if there were no comparative data from freeze-fixed (or Os-FeCN fixed) samples available. Thus, we suppose that tapetum cells remain much longer physiologically and morphologically intact than generally assumed, regardless of species-specific variations. This notion is also

Figs. 5, 6. *Tillandsia pallidoflavens* subjected to high pressure freeze fixation and freeze substitution (same stage as Fig. 4). The tapetum cells show rich cytoplasmic ultrastructure; thus, they appear morphologically (and physiologically) intact in contrast to conventionally chemically fixed samples as shown in Fig. 4. Abundant ER (arrow-heads) is clearly visible. – Fig. 6 shows Golgi bodies with a clear cis-to-trans polarity and prominent post Golgi apparatus structures (PGS). Some of the secretory vesicles (SV) have granular contents. – Fig. 5: bar = 1 μm, Fig. 6: bar = 0,25 μm

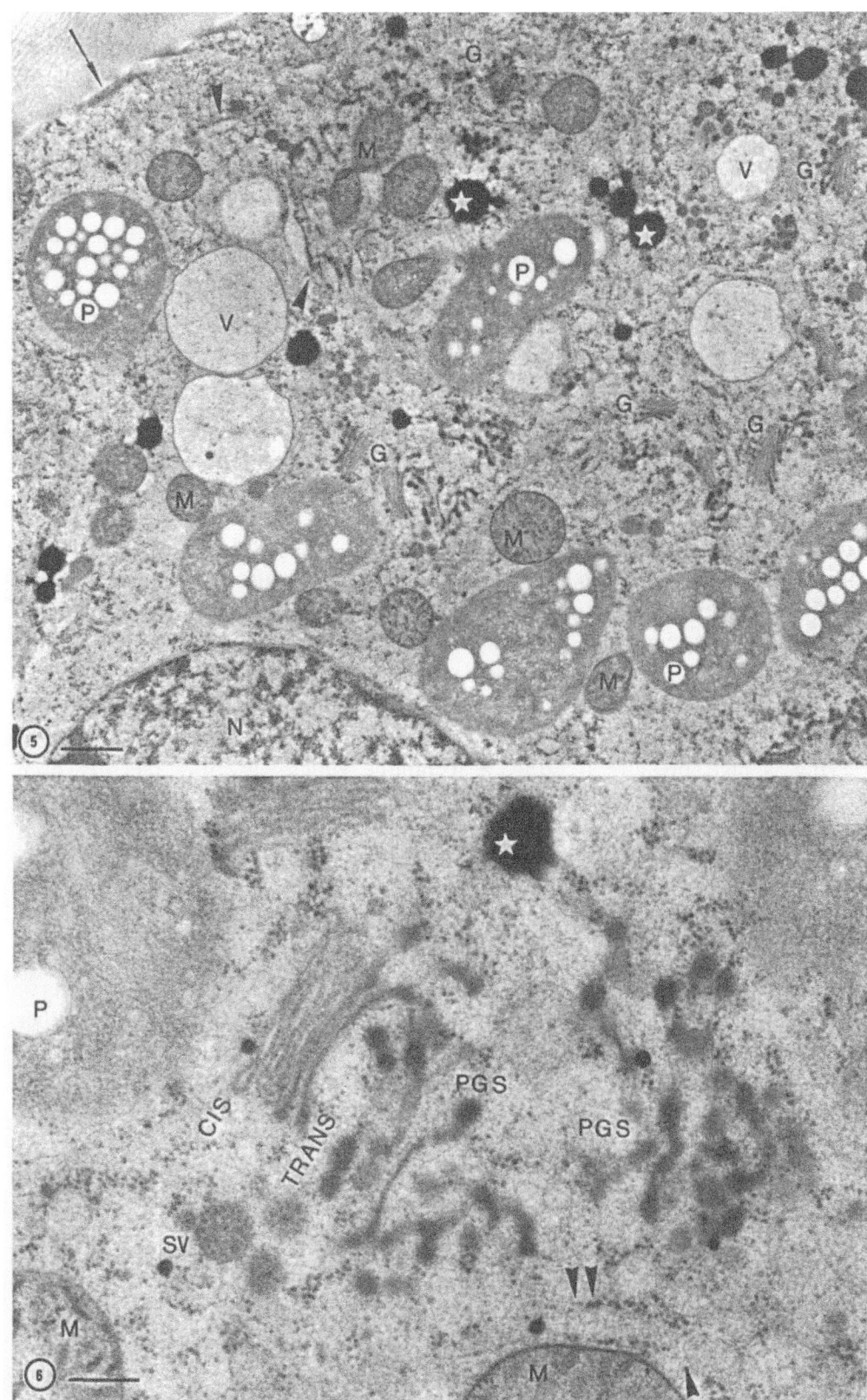

supported by recent data presented by others (Murgia & al. 1991, Weber 1992b, Audran & Dan Dicko-Zafimahova 1992). Terms like "tapetum maturation" or "tapetum senescence" according to Murgia & al. (1991) seem, therefore, more appropriate than "tapetum degeneration", cf. Rowley 1993, Hesse 1993.

Although low-temperature techniques provide an indispensable complementary preparation method some problems and artifacts still remain. Tiwari & Gunning (1986c) have used rapid freeze fixation followed by freeze substitution for an elegant study of tapetum cell surface structures; these techniques allowed to verify that the so-called perispore membrane convolutions of the tapetum in *Canna* are not artificial, i. e. not caused by chemical fixation. For the investigation of cytoplasmic ultrastructure of the tapetum cells, however, rapid freeze fixation seems not to be the freezing technique of choice (Tiwari & Gunning 1986c): When no cryoprotectives are used rapid freezing methods are generally restricted to thin samples, such as single cells, epidermal cells etc. (Robards 1991 and references therein). By contrast, tapetum cells develop within the center of a complex organ, the anther. The multi-layered anther wall usually prevents rapid cryoimmobilization of the tapetum cells which results in freezing artifacts, i. e. ice crystal damage. In order to obtain samples small enough for successful rapid freeze fixation, tapetum cells must either be gently extruded from unfixed anthers (Tiwari & Gunning 1986c), or whole anthers must be cut into small pieces before freezing (Fitzgerald & al. 1993b: this volume). These procedures are problematical, because any wounding of the anther tissues may result in mechanical damage and/or osmotic stress of the (tapetum) cells provoking traumatic reactions prior to freezing. A valuable alternative to rapid freezing is high pressure freezing, a technique which has only recently been introduced (Müller & Moor 1984, Craig & Staehelin 1988). High pressure freezing does neither require extensive specimen preparation nor treatment with cryoprotective chemicals prior to freezing (Studer & al. 1989 and references therein). High pressure freezing permits freeze fixation of large samples up to 0,5 mm thick. It has been successfully applied to freeze intact anthers (Hess 1992, 1993, Staiger & al. 1993) yielding highly reproducible results irrespective of the species investigated (we have tested the method so far on 15 monocot and dicot taxa belonging to 8 and 4 families, respectively). Since the anthers are neither mechanically nor chemically hurt prior to high pressure freeze fixation the risks of inducing artifacts due to traumatic reactions of the tapetum cells are negligible. This is, in our opinion, an indispensable prerequisite for the investigation of any cell types which undergo cytoplasmic degeneration or senescence.

Figs. 7, 8. – Fig. 7. Anther tapetum of *Tilia platyphyllos* prior to microspore mitosis prepared by conventional glutaraldehyde (GA)/osmium fixation. Endomembranes and/or organelles are present but not adequately preserved. The ground cytoplasm as extracted during specimen preparation. Bar = 0,5 μm. – Fig. 8. Anther tapetum of *Tilia platyphyllos* prior to microspore mitosis prepared by glutaraldehyde/Os-FeCN fixation. Ultrastructure preservation is considerably better than in Figs. 1, 2, 4, 7. Organelles labelled as in Figs. 1, 2. Bar = 0,5 μm

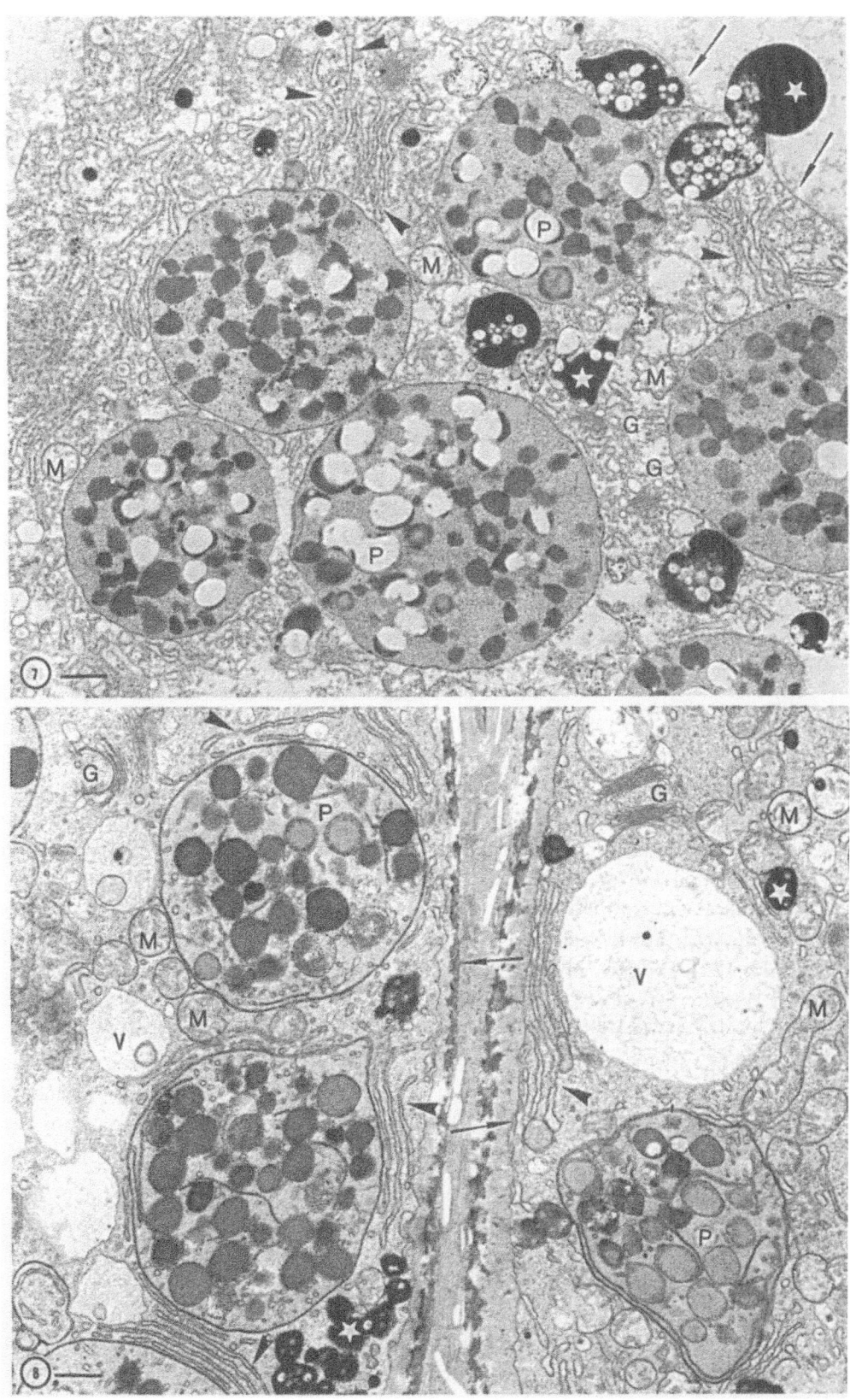

Conclusion. On the basis of information available we presume that high pressure freezing of intact anthers without any chemical cryoprotection is at present the preparation method of choice for the ultrastructural study of tapetum maturation/senescence/degeneration stages.

B) Short review of the recent literature

1) Aspects of primary/secondary metabolisms (lipids, polysaccharides, proteins). Secretory events/pathways including the formation of viscin threads, orbicules/Ubisch-bodies and pollenkitt

Sporopollenin and pollen or orbicular wall formation was investigated by CHEN & al. (1989: *Pinus*), who disagree in some points with ROWLEY & WALLES (1987), further by CLEMENT & AUDRAN (1992, 1993: this volume:), EL-GHAZALY & JENSEN (1990: *Triticum*), FERNANDO & CASS (1993: *Butomus*), FITZGERALD & al. (1993a: *Acacia*) and GABARAYEVA (1990, 1991: *Magnolia*). UEHARA & al. (1991) present data on microspore wall formation in *Isoëtes*. PARKINSON (1988) describes unusual "orbicular projections" in the fern *Psilotum*; other interesting features of sporogenesis in *Psilotum* are the lack of any callosic special wall and the invasion of the plasmodial tapetum into the locule as early as during meiosis (PARKINSON 1987; note: Especially with regard to a better understanding of evolutionary aspects it is a pity that the tapetum in non-spermatophytic taxa, i.e. mosses and ferns, is generally not in the focus of interest, apart from the above mentioned papers). TAKAHASHI & SKVARLA (1990) present a hypothesis on viscin thread formation in *Oenothera*: Branched fibrous strings connecting the ektexine with the tapetum surface are observed immediately after callose dissolution; later, these strings become transformed into viscin threads. GUBATZ & WIERMANN (1992) report on sporopollenin biosynthesis; the same authors also give a comprehensive review on "Pollen wall and sporopollenin" (WIERMANN & GUBATZ 1992). The biochemistry of pollen pigmentation is discussed by BEERHUES & al. (1993: this volume).

Another important secondary product is pollenkitt. The nearly dogmatic view that pollenkitt is exclusively synthesized by only one kind of tapetal organelle has been abandoned. Many authors had already been claiming that pollenkitt is not merely derived from the tapetal plastids, but subcellular localization of a second source was not yet available. There is now considerable, sufficient evidence from various point of views, from various taxa and supported by different techniques that the plastids/elaioplasts (see also PACINI & al. 1992) and the endoplasmic reticulum both contribute to pollenkitt formation: The ER-derived pollenkitt precursors are already produced during early tapetum development (i. e. from the free microspore stage onwards) whereas the plastidial components in general start accumulating prior to microspore mitosis (WEBER 1991, 1992ab, HESSE 1993: this volume, herein outstanding literature, HESS & HESSE 1993). ROWLEY & EL GHAZALY (1992) compare different fixation protocols for TEM presenting remarkably improved preservation of lipophilic substances, in particular pollenkitt. JORDAAN & KRUGER (1993) report preliminary results on the influence of malachite green on pollenkitt preservation. HESS & HESSE (1993) discuss some aspects of lipid preservation by freeze substitution protocols. Further data on the lipid metabolism of anther tissues are provided by elegant light microscopic studies. EVANS & al. (1992) treated cryostat sections of fresh, unfixed *Brassica* anthers with different fluorochromes. Quantitative micro-fluorometric analysis revealed that the accumulation of lipid bodies within the tapetum strictly coincides with high levels of acyl carrier protein, a protein integrally involved in lipid synthesis. NOHER DE HALAC & al. (1992) report changes in lipid and polysaccharide contents within different tissues during anther development of *Oenothera*; their LM micrographs (obtained from fixed and resin embedded samples) also demonstrate some unusual staining patterns of the callosic special wall of meiocytes and microspore tetrads. Further data on tapetal polysaccharide meta-bolism and/or secretion are reported by HESS & HESSE (1993) on *Ledebouria*.

Due to the use of freeze fixation and freeze substitution FITZGERALD & al. (1993b: this volume) could show considerable amounts of proteins accumulating at the surface of the tapetum cells in the orchid *Pterostylis*. WANG & al. (1992a) used gel electrophoresis for the analysis of floral organ specific and floral enriched proteins in *Lilium* anthers. The patterns of protein accumulation as revealed by this technique correlate with distinct periods of microspore and/or tapetum development. The same authors also prepared antibodies for the *in situ* localization of tapetum specific proteins in *Lilium* (WANG & al. 1992b); their data obtained from LM examination of immunogold labelled glycol methacrylate sections showed a strict coincidence of the maximal protein level within the tapetum cells and the peak of tapetum secretory activities. On the TEM-level STAIGER & al. (1993) yielded highly specific immuno-gold labelling of tapetum specific proteins in *Sinapis* (see also below: STAIGER & APEL 1993). These proteins are localized within the forming exine and tapetum derived sporopolleninous by-products during the free microspore stage (i. e. the forming peritapetal membrane and orbicule-like globules within the anther locule). The molecular characterization of other tapetum specific proteins gave rise to the idea that some of these proteins are released into the anther locule either directly or following storage in the tapetum vacuoles (KOLTUNOW & al. 1990, NACKEN & al. 1991, PAUL & al. 1992); data on the subcellular localization are not yet available.

2) Regulation of tapetum and microspore development. Abnormal tapetum/microspore development

Cytoplasmic channels are formed after meiosis between the tapetal cells in *Zea mays* (PERDUE & al. 1992) and *Acacia* (FITZGERALD & al. 1993a); they presumably provide the mechanism by which differentiation of the tapetum cells is synchronized within the anther (PERDUE & al. 1992). Microspore growth in *Centrolepis* is suspected to be regulated and synchronized by the uptake of tapetum derived nutrients via exine spines of the microspores intimately associated with the tapetum cytoplasm (ROWLEY & DUNBAR 1990). TIRLAPUR & WILLEMSE (1992) fluorimetrically determined the patterns of membrane associated calcium and calmodulin within the anthers of *Gasteria* (see also WILLEMSE 1993: this volume); the role of tapetum-originated calcium in callose synthesis is discussed and transports of calcium from the tapetum to the young pollen grain are postulated. DICKINSON (1992) has reviewed recent data on gene expression during meiosis and microspore development as well as the role of the tapetum in regulating these processes (note: the paper includes comments on studies still in progress). Further data on spatial and temporal patterns of tapetum specific gene expression have been reported from KOLTUNOW & al. (1990), SMITH &

al. (1990), Scott & al. (1991), Toriyama & al. (1991), Foster & al. (1992) and Staiger & Apel (1993). An interesting method for non-radioactive labeling of total mRNA in plastic embedded anther sections was introduced by Chandra Sekhar & Williams (1992); data on mRNA levels during microsporogenesis were presented.

Any failure in tapetum control results in pollen grain malfunction and sterility (Bino 1985, Kaul 1988). Inducing male sterility by genetic engeneering gave for the first time direct evidence of the key role of the tapetum in the regulation of normal pollen development (Koltunow & al. 1990, Mariani & al. 1990, 1992, van der Meer & al. 1992, Worrall & al. 1992, Aart & al. 1993, De Block & Debrouwer 1993). For instance, van der Meer & al. (1992) demonstrated that blocking the flavonoid biosynthesis effects male fertility. Worrall & al. (1992) showed that premature callose dissolution causes the same male sterility aspects in an genetically engineered *Nicotiana* line as it does in a naturally sterile *Petunia* line. Another type of experimental inhibition of normal microgametogenesis includes the influence of high and/or low cultivation temperatures (Polowick & Sawhney 1987, 1990, 1991) and the effects of chemical agents (Cross & Ladyman 1991, Schulz & al. 1993) on tapetum and/or pollen development; the latter study on *Triticum* reports on tapetum hypertrophy prior to abnormal tapetum degeneration and irregular sporopollenin deposition patterns induced by phenylcinnoline carboxylates (with regard to alterations of the sporopollenin secretory pathway we would like to refer once more to the classical paper by Tiwari & Gunning 1986b). Of course, most papers deal with the induction of male sterility. Using genetic engineering the reverse is also possible: Mariani & al. (1992) showed that suppressing the cytotoxic ribonuclease activity by forming cell-specific RNase/RNase inhibitor complexes restores male fertility in *Brassica*. Nevertheless, we are, at present, still far away from understanding the causal relationships of normal/abnormal sporo- and gameto-genesis, despite much progress in this field.

Further papers comparing fertile with sterile lines are: Grant & al. (1986) report that microspore degeneration in sterile *Brassica* starts very early, i. e. during the tetrad stage, but the causal relationship between the observed tapetum proliferation and/or hypertrophy and the deletion of microspores within the callose remains obscure. Similar observations on delayed tapetum senescence were made by Sawhney & Bhadula (1988); their LM study also revealed abnormal features of sporopollenin deposition. Halldén & al. (1991) describe unusual features in sterile *Beta* tapetum cells (e.g. proliferation of tapetal cells and wall dissolutions in early stages, cf. in this point the paper by Rowley 1993: this volume!). Holford & al. (1991) report on three different types of abnormal tapetum behaviour causing male sterility in *Allium* (in type 1 the tapetum degenerates extremely early, in type 2 the tapetum undergoes hypertrophy and autolyzation, and, interestingly, in type 3 the tapetum remains structurally intact for an abnormally long time: this proves once more that exact synchronization between tapetum and pollen development is absolutely necessary). Moussel & al. (1990ab, 1992) continue their series of papers on *Vicia*: virus-like particles may contribute to male sterility. Majewska-Sawka & al. (1993) show that abnormal tapetum features in *Beta* occur as early as during the tetrad stage and that orbicule formation is hindered by peculiar tapetum dysfunctions; the authors further describe volume reductions of mitochondria as a cytological feature of sterility. Impeded mitochondria multiplication is assumed to cause male sterility in *Nicotiana*, too (POLLAK 1992); in contrast, HOLFORD & al. (1991) found no differences in mitochondrial volumes in their type-2-male sterile *Allium*. This demonstrates once more the manifoldness of sterility features and putative reasons. NOHER DE HALAC & al. (1990) observed that in male sterile lines of *Oenothera* the ectexine is not formed and that the endexine is presumably dissolved by enzymes. Rounding off this paragraph, HORNER & al. (1993) present an interesting methodology for the investigation of, so far neglected, cytological features of tapetum protoplasts of cytoplasmic male sterile plants employing squash preparations, tapetal protoplast separation via flow cytometry, image analysis, and electron microscopy.

3) Concluding notes on tapetum typology

In the light of the recently achieved optimal structural preservation of the delicate tapetum cells especially in their secretory phase the following points should be considered. The distinction between the two main tapetum types, i.e. cellular/ secretory or parietal type, and the amoeboid or periplasmodial type, has become less evident (see also TIWARI & GUNNING 1986ab). The conversion of a tapetum with "secretory" features to one with some "amoeboid" characters can take place as early as in the tetrad stage. Thus, even during early developmental stages it is sometimes impossible to make distinct classifications; during later stages the differences become blurred more and more as already pointed out by PACINI & FRANCHI (1991). To further complicate matters new subtypes and/or transitional types have recently been found (TIWARI & GUNNING 1986ab, PACINI & KEIJZER 1989, PACINI 1990, AUDRAN & DAN DICKO-ZAFIMAHOVA 1992). Detailed studies of tapetum features show that five or even more periods in tapetum activities can be detected (e. g. MOUSSEL & al. 1992), and/or cycles of tapetal activities occur (e. g. KRONESTEDT-ROBARDS & ROWLEY 1989, ROWLEY & al. 1992, and ROWLEY 1993: this volume). Moreover, the amoeboid tapetum of pteridophytes and angiosperms is an example for evolutionary convergence. Thus, the reconstruction of the phylogenesis as well as the functional interpretation of evolutionary trends of the tapeta in the Embryophyta cannot be based solely on typological characters. PACINI & al. (1985) avoided any typology. According to these authors there exist three evolutionary trends in Spermatophyta: 1. an intrusion of tapetum cells into the loculus, 2. the loss of tapetal cell walls, and 3. a better nutrition through direct contact with the spores in narrow anthers. Nevertheless, the terms "secretory = parietal" and "amoeboid = periplasmodial" may be of use for a simplified classification.

We thank Dr. M. MÜLLER (Institut für Zellbiologie, ETH-Zürich) for the kind permission to use a Balzers high pressure freezer at his laboratory. We are indebted to Mag. M. G. SCHLAG (Institut für Botanik, Universität Wien) for helpful discussion. Thanks are also due to U. SCHACHNER (Institut für Botanik, Universität Wien) for excellent technical assistance. This work was supported by the Austrian "Fonds zur Förderung wissenschaftlicher Forschung" (P 8138 BIO). Part of this work was presented at the 8th International Palynological Congress held in September 1992 in Aix-en-Provence/France.

References

Aarts, M. G. M., Dirkse, W. G., Stiekema, W. J., Pereira, A., 1993: Transposon tagging of a male sterility gene in *Arabidopsis*. – Nature **363**: 715–717.

Albertini, L., Souvré, A., Audran, J. C., 1987: Le tapis de l'anthère et ses relations avec les microsporocytes et les grains de pollen. – Rev. Cytol. Biol. végét. – Bot. **10**: 211–242.

Audran, J. C., Dan Dicko-Zafimahova, L., 1992: Aspects ultrastructuraux et cytochimiques du tapis staminal chez *Calotropis procera* (*Asclepiadaceae*). – Grana **31**: 253–272.

Beerhues, L., Rittscher, M., Schöpker, H., Schwerdtfeger, C., Wiermann, R., 1993: The significance of the anther tapetum in the biochemistry of pollen pigmentation – an overview. – Pl. Syst. Evol. [Suppl.] **7**: 117–125.

Bhandari, N. N., 1984: The Microsporangium. – In Johri, B. M. (Ed): Embryology of Angiosperms, pp. 53–121. – Berlin, Heidelberg, New York, Tokyo: Springer.

Bino, R. J., 1985: Histological aspects of microsporogenesis in fertile, cytoplasmic male sterile and restored fertile *Petunia hybrida*. – Theor. Appl. Genet. **69**: 423–428.

Carniel, K., 1963: Das Antherentapetum. – Österr. Bot. Z. **110**: 145–176.

Chandra Sekhar, K. N., Williams, E. G., 1992: Nonradioactive in situ localization of poly(A) + RNA during pollen development in anthers of tobacco (*Nicotiana tabacum* L.). – Protoplasma **169**: 9–17.

Chapman, G. P., 1987: The tapetum. – Int. Rev. Cytol. **107**: 111–125.

Chen, Z.-K., Wang, F. H., Zhou, F., 1989: The origin, development and fine structure of Ubisch bodies in *Pinus bungeana*. – Cathaya **1**: 109–118.

Clément, C., Audran, J.-C., 1992: Apports de la cytochimie à la connaissance des orbicules dans l'anthère de *Lilium* (Liliacées). 1 – Le coeur orbiculaire. – Bull. Soc. bot. Fr. **139**, Lettres bot.: 369–376.

Clément, C., Audran, J.-C., 1993: Cytochemical and ultrastructural evolution of orbicules in *Lilium*. – Pl. Syst. Evol. [Suppl.] **7**: 63–74.

Craig, S., Staehelin, L. A., 1988: High pressure freezing of intact plant tissues. Evaluation and characterization of novel features of the endoplasmic reticulum and associated membrane systems. – Eur. J. Cell. Biol. **46**: 80–93.

Cross, J. W., Ladyman, J. A. R., 1991: Chemical agents that inhibit pollen development: tools for research. – Sex. Plant Reprod. **4**: 235–243.

De Block, M., Debrouwer, D., 1993: Engeneered fertility control in transgenic *Brassica napus* L.: Histochemical analysis of anther development. – Planta **189**: 218–225.

Dickinson, H. G., 1992: Microspore derived embryogenesis. – In Cresti, M., Tiezzi, A., (Eds): Sexual plant reproduction, pp. 1–15. – Berlin, Heidelberg, New York, London, Paris, Tokyo, Hong Kong, Barcelona, Budapest: Springer.

Echlin, P., 1971: The role of the tapetum during microsporogenesis of angiosperms. – In Heslop-Harrison, J., (Ed): Pollen: Development and Physiology, pp. 41–61. – London: Butterworths.

El-Ghazaly, G., Jensen, W. A., 1990: Development of wheat (*Triticum aestivum*) pollen wall before and after effect of a gametocide. – Can. J. Bot. **68**: 2509–2516.

Evans, D. E., Taylor P. E., Singh, M. B., Knox, R. B., 1992: The interrelationship between the accumulation of lipids, protein and the level of acyl carrier protein during the development of *Brassica napus* L. pollen. – Planta **186**: 343–354.

Fernando, D. D., Cass, D. D., 1993: The role of plasmodial tapetum in pollen wall development: pathways of sporopollenin secretion in *Butomus umbellatus* L. – Amer. J. Bot. **80**, Suppl.: Abstracts 66.

Fitzgerald, M. A., Calder, D. M., Knox, R. B., 1993a: Character states of development and initiation of cohesion between compound pollen grains of *Acacia paradoxa*. – Ann. Bot. **71**: 51–59.

FITZGERALD, M. A., CALDER, D. M., KNOX, R. B., 1993b: Secretory events in the freeze-substituted tapetum of the orchid *Pterostylis concinna*. – Pl. Syst. Evol. [Suppl.] **7**: 53–62

GABARAYEVA, N., 1990: On the site of sporopollenin precursors synthesis in the developing pollen grains of members of the *Magnoliaceae* family. – Bot. Zh. **75**: 783–791.

GABARAYEVA, N., 1991: The ultrastructure and development of exine and orbicules of *Magnolia delavayi* (*Magnoliaceae*) in the tetrad and the beginning of posttetrad periods. – Bot. Zh. **76**: 10–19.

GRANT, I., BEVERSDORF, W. D., PETERSON, R. L., 1986: A comparative light and electron microscope study of microspore and tapetal development in male fertile and cytoplasmic male sterility oilseed rape (*Brassica napus*). – Can. J. Bot. **64**: 1055–1068.

GUBATZ, S., WIERMANN, R., 1992: Studies on sporopollenin biosynthesis in *Tulipa* anthers. III. Incorporation of specifically labeled ^{14}C-phenylalanine in comparison to other precursors. – Bot. Acta **105**: 407–413.

HALLDÉN, C., KARLSSON, G., LIND, C., MOLLER, I. M., HENEEN, W. K., 1991: Microsporogenesis and tapetal development in fertile and cytoplasmic male-sterile sugar beet (*Beta vulgaris* L.). – Sex. Plant Reprod. **4**: 215–225.

HEPLER, P. K., 1981: The structure of the endoplasmic reticulum revealed by osmium tetroxide-potassium ferricyanide staining. – Eur. J. Cell Biol. **26**: 102–110.

HESS, M. W., 1992: High-pressure freezing/freeze-substitution of delicate reproductive tissues in plants. – In MEGÍAS-MEGÍAS, L., RODRÍGUEZ-GARCÍA, M. I., RÍOS, A., ARIAS, J. M. (Eds): Electron Microscopy 92, Proceedings of the 10th European Congress on Electron Microscopy held in Granada, Spain, 7–11 September 1992, Volume III: Biological Sciences. pp. 67–68. – Granada: Secretariado de Publicaciones de la Universidad de Granada.

HESS, M. W., 1993: Cell-wall development in freeze-fixed pollen: Intine formation of *Ledebouria socialis* (*Hyacinthaceae*). – Planta **189**: 139–149.

HESS, M. W., GLASER, A., 1993: A simple and inexpensive device for freeze substitution at 183 K/-90C. – Biotech. Histochem. (in press).

HESS, M. W., HESSE, M., 1993: Ultrastructural observations on anther tapetum development in freeze fixed *Ledebouria socialis* (*Hyacinthaceae*). – Planta (in press).

HESSE, M., 1991: Cytology and morphogenesis of pollen and spores. – Prog Bot. **52**: 19–34.

HESSE, M., 1993: Pollenkitt development and composition in *Tilia platyphyllos* (*Tiliaceae*) analysed by conventional and energy filtering TEM. – Pl. Syst. Evol. [Suppl.] **7**: 39–52.

HOLFORD, P., CROFT, J., NEWBURY, H. J., 1991: Structural studies of microsporogenesis in fertile and male-sterile onions (*Allium cepa* L.) containing the cms-S cytoplasm. – Theor. Appl. Genet. **82**: 745–755.

HORNER, H. T., HALL, V. L., VARGAS OLVERA, M. A., 1993: Isolation, sorting, and characterization of uninucleate and binucleate tapetal protoplasts from anthers of normal and texas cytoplasmic male-sterile *Zea mays* L. – Protoplasma **173**: 48–57.

JORDAAN, A., KRUGER, H., 1993: Pollen wall ontogeny of *Felicia muricata* (*Asteraceae: Astereae*). – Ann. Bot. **71**: 97–105.

KAMINSKYJ, S. G. W., JACKSON, S. L., HEATH, I. B., 1992: Fixation induces differential polarized translocations of organelles in hyphae of *Saprolegnia ferax*. – J. Microsc. **167**: 153–168.

KAUL, M. L. K., 1988: Male sterility in higher plants. – In FRANKEL, R., GROSSMAN, M., LINSKENS, H. F., MALIGA, P., RILEY, R. (Eds): Monographs on theoretical and applied genetics. Volume **10**, pp. 15–95. – Berlin, Heidelberg, New York: Springer.

KOLTUNOW, A. M., TRUETTNER, J., COX, K. H., WALLROTH, M., GOLDBERG, R. B., 1990:

Different temporal and spatial gene expression patterns occur during anther development. – The Plant Cell **2**: 1201–1224.

Kronestedt-Robards, E. C., Rowley, J. R., 1989: Pollen grain development and tapetal changes in *Strelitzia reginae* (*Strelitziaceae*). – Amer. J. Bot. **76**: 856–870.

Majewska-Sawka, A., Rodriguez-Garcia, M. I., Nakashima, H., Jassem, B., 1993: Ultrastructural expression of cytoplasmic male sterility in sugar beet (*Beta vulgaris* L.). – Sex. Plant Reprod. **6**: 22–32.

Mariani, C., De Beuckeleer, M., Truettner, J., Leemans, J., Goldberg, R. B., 1990: Induction of male sterility in plants by a chimaeric ribonuclease gene. – Nature **347**: 737–741.

Mariani, C., Gossele, V., De Beuckeleer, M., De Block, M., Goldberg, R. B., De Greef, W., Leemans, J., 1992: A chimaeric ribonuclease-inhibitor gene restores fertility to male sterile plants. – Nature **357**: 384–387.

Mineyuki, Y., Gunning, B. E. S., 1988: Streak time-lapse video microscopy: analysis of protoplasmic motility and cell division in *Tradescantia* stamen hair cells. – J. Microsc. **150**: 41–55.

Moussel, B., Moussel, C., Audran, J. C., 1990a: Nucleo-cytoplasmic male sterility in faba bean (*Vicia faba* L.). A cytological overview. – Phytomorphology **40**: 69–77.

Moussel, B., Moussel, C., Duc, G., Audran, J. C., 1990b: Modalités cytologiques de l'avortement des microspores ou du pollen chez plusieurs lignées mâle-stériles de Féverole (*Vicia faba* L.) (stérilité mâle nucléo-cytoplasmique). – Bull. Soc. bot. Fr. **137**, Actual. bot.: 57–64.

Moussel, B., Moussel, C., Audran, J. C., 1992: La stérilité mâle nucléo-cytoplasmique chez la féverole (*Vicia faba* L.). IX. – Grana **31**: 25–48.

Müller, M., Moor, H., 1984: Cryofixation of thick specimens by high pressure freezing. – In Revel, J. P., Barnard, T., Haggins, G. H. (Eds): The science of biological specimen preparation. pp. 131–138. – Chicago: SEM Inc., AMF O'Hare.

Murgia, M., Charzynska, M., Rougier, M., Cresti, M., 1991: Secretory tapetum of *Brassica oleracea* L.: polarity and ultrastructural features. – Sex. Plant Reprod. **4**: 28–35.

Nacken, W. K. F., Huijser, P., Beltran, J. P., Saedler, H., Sommer, H., 1991: Molecular characterization of two stamen-specific genes, *tap1* und *fil1*, that are expressed in the wild type, but not in the *deficiens* mutant of *Antirrhinum majus*. – Mol. Gen. Genet. **229**: 129–136.

Noher de Halac, I., Cismondi, I. A., Harte, C., 1990: Pollen ontogenesis in *Oenothera*: a comparison of genotypically normal anthers with the male-sterile mutant *sterilis*. – Sex. Plant Reprod. **3**: 41–53.

Noher, de Halac, I., Fama, G., Cismondi, I. A., 1992: Changes in lipids and polysaccharides during pollen ontogeny in *Oenothera* anthers. – Sex. Plant Reprod. **5**: 110–116.

Pacini, E., 1990: Tapetum and microspore function. – In Blackmore, S., Knox, R. B., (Eds): Microspores. Evolution and ontogeny, pp. 213–237. – London, San Diego, New York, Boston, Sydney, Tokyo, Toronto: Academic Press.

Pacini, E., Franchi, G. G., 1991: Diversification and evolution of the tapetum. – In Blackmore, S., Barnes, S. H., (Eds): Pollen and Spores. Patterns of Diversification, pp. 301–316. – Oxford: Clarendon Press.

Pacini, E., Franchi, G. B., 1993: Role of tapetum in pollen and spore dispersal. – Pl. Syst. Evol. [Suppl.] **7**: 1–11.

Pacini, E., Keijzer, C. J., 1989: Ontogeny of intruding non-periplasmodial tapetum in the wild chicory, *Cichorium intybus* (*Compositae*). – Pl. Syst. Evol. **167**: 149–164.

Pacini, E., Franchi, G. G., Hesse, M., 1985: The tapetum: its form, function, and possible phylogeny in *Embryophyta*. – Pl. Syst. Evol. **149**: 155–185.

PACINI, E., TAYLOR, P. E., SINGH, M. B., KNOX, R. B., 1992: Development of plastids in pollen and tapetum of Rye-Grass, *Lolium perenne* L. - Ann. Bot. **70**: 179–188.

PARKINSON, B. M., 1987: Tapetal organization during sporogenesis in *Psilotum nudum*. - Ann. Bot. **60**: 353–360.

PARKINSON, B. M., 1988: A tapetal membrane in *Psilotum nudum* (L.) BEAUV. - Ann. Bot. **61**: 695–703.

PAUL, W., HODGE, R., SMARTT, S., DRAPER, J., SCOTT, R., 1992: The isolation and characterization of the tapetum-specific *Arabidopsis thaliana* A9 gene. - Plant Molec. Biol. **19**: 611–622.

PERDUE, T. D., LOUKIDES, C. A., BEDINGER, P. A., 1992: The formation of cytoplasmic channels between tapetal cells in *Zea mays*. - Protoplasma **171**: 75–79.

POLLAK, P. E., 1992: Cytological differences between a cytoplasmic male sterility tobacco cybrid and its fertile counterpart during early development. - Amer. J. Bot. **79**: 937–945.

POLOWICK, P. L., SAWHNEY, V. K., 1987: A scanning electron microscopic study on the influence of temperature on the expression of cytoplasmic male sterility in *Brassica napus*. - Can. J. Bot. **65**: 807–814.

POLOWICK, P. L., SAWHNEY, V. K., 1990: Microsporogenesis in a normal line and in the *ogu* cytoplasmic male-sterile line of *Brassica napus*. - Sex. Plant Reprod. **3**: 263–276.

POLOWICK, P. L., SAWHNEY, V. K., 1991: Microsporogenesis in a normal line and in the *ogu* cytoplasmic male sterility line of *Brassica napus:* II. The influence of intermediate and low temperatures. - Sex. Plant Reprod. **4**: 22–27.

ROBARDS, A. W., 1991: Rapid-freezing methods and their applications. - In HALL, J. L., HAWES, C., (Eds): Electron microscopy of plant cells, pp. 257–312. - London, San Diego, New York, Boston, Sydney, Tokyo, Toronto: Academic Press.

ROBARDS, A. W., STARK, M., 1988: Nectar secretion in *Abutilon*: a new model. - Protoplasma **142**: 79–91.

ROWLEY, J. R., 1993: Cycles of hyperactivity in tapetal cells. - Pl. Syst. Evol. [Suppl.] **7**: 23–37.

ROWLEY, J. R., DUNBAR, A., 1990: Outward extension of spinules in exine of *Centrolepis aristata* (*Centrolepidaceae*). - Bot. Acta **103**: 355–359.

ROWLEY, J. R., EL-GHAZALY, G., 1992: Lipid in wall and cytoplasm of *Solidago* pollen. - Grana **31**: 273–283.

ROWLEY, J. R., WALLES, B., 1987: Origin and structure of Ubisch bodies in *Pinus sylvestris*. - Acta Soc. Bot. Pol. **56**: 215–227.

ROWLEY, J. R., WALLES, B., 1993: Cell differentiation in microsporangia of *Pinus sylvestris*. V. Diakinesis to tetrad formation. - Nord. J. Bot. **13**: 67–82.

ROWLEY, J. R., GABARAYEVA, N. I., WALLES, B., 1992: Cyclic invasion of tapetal cells into loculi during microspore development in *Nymphaea colorata* (*Nymphaceae*). - Amer. J. Bot. **79**: 801–808.

SAWHNEY, V. K., BHADULA, S. K., 1988: Microsporogenesis in the normal and male-sterile stamenless-2 mutant of tomato (*Lycopersicon esculentum*). - Can. J. Bot. **66**: 2013–2021.

SCHULZ, P. J., CROSS, J. W., ALMEIDA, E., 1993: Chemical agents that inhibit pollen development: effects of the phenylcinnoline carboxylates SC-1058 and SC-1271 on the ultrastructure of developing wheat anthers (*Triticum aestivum* L. var Yecora rojò). - Sex. Plant Reprod. **6**: 108–212.

SCOTT, R., DAGLESS, E., HODGE, R., PAUL, W., SOUFLERI, I., DRAPER, J., 1991: Patterns of gene expression in developing anthers of *Brassica napus*. - Plant Molec. Biol. **17**: 195–207.

SHIVANNA, K. R., JOHRI, B. M., 1985: The angiosperm pollen. Structure and function. - New York, Chichester, Brisbane, Toronto, Singapore: John Wiley & Sons.

Smith, A. G., Gasser, C. S., Budelier, K., Fraley, R., 1990: Identification and characterization of stamen- and tapetum-specific genes from tomato. - Mol. Gen. Genet. **222**: 9–16.

Staiger, D., Apel, K., 1993: Molecular characterization of two cDNAs from *Sinapis alba* expressed specifically at an early stage of tapetum development. - Plant J. (in press).

Staiger, D., Kappeler, S., Müller, M., Apel, K., 1993: The proteins encoded by two tapetum-specific transcripts, *Satap35* and *Satap44*, from *Sinapis alba* L. are localized in the exine cell wall layer of developing microspores. - Planta (in press).

Studer, D., Michel, M., Müller, M., 1989: High pressure freezing comes of age. - Scanning Microsc. Suppl. **3**: 253–269.

Takahashi, M., Skvarla, J. J., 1990: Pollen development in *Oenothera biennis* (*Onagraceae*). - Amer. J. Bot. **77**: 1142–1148.

Telmer, C. A., Newcomb, W., Simmonds, D. H., 1993: Microspore development in *Brassica napus* and the effect of high temperature on division in vivo and in vitro. - Protoplasma **172**: 154–165.

Tirlapur, U. K., Willemse, M. T. M., 1992: Changes in calcium and calmodulin levels during microsporogenesis, pollen development and germination in *Gasteria verrucosa* (Mill.) H. Duval. - Sex. Plant Reprod. **5**: 214–223.

Tiwari, S. C., Gunning, B. E. S., 1986a: Development of tapetum and microspores in *Canna* L.: an example of an invasive but non-syncytial tapetum. - Ann. Bot. **57**: 557–563.

Tiwari, S. C., Gunning, B. E. S., 1986b: Colchicine inhibits plasmodium formation and disrupts pathways of sporopollenin secretion in the anther tapetum of *Tradescantia virginiana* L. - Protoplasma **133**: 115–128.

Tiwari, S. C., Gunning, B. E. S., 1986c: Development and cell surface of a non-syncytial invasive tapetum in *Canna*: ultrastructural, freeze-substitution, cytochemical and immunofluorescence study. - Protoplasma **134**: 1–16.

Toriyama, K., Thorsness, M. K., Nasrallah, J. B., Nasrallah, M. E., 1991: A *Brassica* S locus gene promoter directs sporophytic expression in the anther tapetum of transgenic *Arabidopsis*. - Dev. Biol. **143**: 427–431.

Uehara, K., Kurita, S., Sahashi, N., Ohmoto, T., 1991: Ultrastructural study on microspore wall morphogenesis in *Isoëtes japonica* (*Isoëtaceae*). - Amer. J. Bot. **78**: 1182–1190.

Van der Meer, I. M., Stam, M. E., Van Tunen, A. J., Mol, J. N. M., Stuitje, A. R., 1992: Inhibition of flavonoid biosynthesis in *Petunia* anthers by antisense RNA: a novel way to engineer nuclear male sterility. - In Ottaviano, E., Mulcahy, D. L., Sari Gorla, M., Mulcahy, G. B. (Eds): Angiosperm pollen and ovules, pp. 22–27. - New York, Berlin, Heidelberg, London, Paris, Tokyo, Hong Kong, Barcelona, Budapest: Springer.

Wang, C.-S., Walling, L. L., Eckard, K. J., Lord, E. M., 1992a: Patterns of protein accumulation in developing anthers of *Lilium longiflorum* correlate with histological events. - Amer. J. Bot. **79**: 118–127.

Wang, C.-S., Walling, L. L., Eckard, K. J., Lord, E. M., 1992b: Immunological characterization of a tapetal protein in developing anthers of *Lilium longiflorum*. - Plant Physiol. **99**: 822–829.

Weber, M., 1991: The transfer of pollenkitt in *Smyrnium perfoliatum* (*Apiaceae*). - Ann. Bot. **68**: 63–68.

Weber, M., 1992a: Nature and distribution of the exine-held material in mature pollen grains of *Apium nodiflorum* L. (*Apiaceae*). - Grana **31**: 17–24.

Weber, M., 1992b: The formation of pollenkitt in *Apium nodiflorum* (*Apiaceae*). - Ann. Bot. **70**: 573–577.

Wiermann, R., Gubatz, S., 1992: Pollen wall and sporopollenin. - Int. Rev. Cytol. **140**: 35–72.

WILLEMSE, M. T. M., 1993: Calcium and calmodulin distribution in the tapetum of *Gasteria verrucosa* during anther development. – Pl. Syst. Evol. [Suppl.] **7**: 107–116.

WILSON, T. P., CANNY, M. J., McCULLY, M. E., LEFKOVITCH, L. P., 1990: Breakdown of cytoplasmic vacuoles. A model of endoplasmic membrane rearrangement. – Protoplasma **155**: 144–152.

WORRALL, D., HIRD, D. L., HODGE, R., PAUL, W., DRAPER, J., SCOTT, R., 1992: Premature dissolution of the microsporocyte callose wall causes male sterility in transgenic tobacco. – Plant Cell **4**: 759–771.

Adress of authors: Prof. Dr. MICHAEL HESSE and Dr. MICHAEL W. HESS, Institut für Botanik und Botanischer Garten, Universität Wien, Rennweg 14, A-1030 Wien, Austria.

Conspectus/Outlook

The study of anther tapetum features includes many different aspects of research, not only in the field of botany *sensu stricto* but also in more applied disciplines. The fields of interest in tapetum research are listed below. They range from Basic Research with important topics, to the other pole Applied Research, with fields of interest especially in economy; between both poles a broad transition zone exists, and many topics overlap. The number and diversity of these topics clearly demonstrate the importance of tapetum research.

This volume presents twelve contributions on tapetum biology – generally belonging to basic research. Find below the authors' name (in brackets) of the articles fitting *grosso modo* to the cited topics; the thematic overlapping in most papers had to be neglected.

BASIC RESEARCH
    Methodology (HESSE & HESS)
    Molecular biology (RASOLONJATOVO & al.)
    Cytochemistry (BEERHUES & al., TESTILLANO & al.)
    Cytophysiology (JARDINAUD & al., WILLEMSE)
    Plant cytology (HESSE, ROWLEY)
    Morphology (CLÉMENT & AUDRAN)
    Development (CIAMPOLINI & al., FITZGERALD & al.)
    Taxonomy (−)
    Phylogeny (−)
    Plant and animal ecology (PACINI & FRANCHI)

APPLIED RESEARCH
    Plant reproduction
    Natural or induced male sterility
    Incompatibility
    Plant breeding
    Plant genetics
    Genetic engineering
    Aerobiology
    Allergology

Practically no contribution fits clearly to any topic of applied research (male sterility or genetics are only marginally touched on in the contributions by WILLEMSE, or RASOLONJATOVO & al., or TESTILLANO & al.). As we can see interesting aspects of tapetum biology could not be covered by the Aix-en-Provence meeting. For various reasons applied research was *a priori* underrepresented during the tapetum conference with only two contributions; very unfortunately both contributions – one on engineered tapetum and the other on gene expression, – could not be included in the present volume. This would have been of interest giving more stimulation and promotion to the "classical" tapetum topics; a meeting covering especially such topics should be a goal for the future.

Last but not least a further aspect should be considered: Related tissues with similar nutritive features and functions, as e.g. the embryo suspensor, the endosperm etc., have received even less attention in the past than the anther tapetum. We hope that our example will also stimulate meetings covering such topics.

Subject index

Peter F. Yeo

Secondary Pollen Presentation

Form, Function and Evolution

1993. 55 figures. VIII, 268 pages.
Cloth DM 220,–, öS 1540,–
Reduced price for subscribers to "Plant Systematics and Evolution":
Cloth DM 198,–, öS 1386,–
ISBN 3-211-82448-0

(Plant Systematics and Evolution / Supplementum 6)

Prices are subject to change without notice.

Secondary pollen presentation is presentation of pollen to vectors by structures other than anthers, either passively or via a specialized protection and delivery system. The main part of the book describes secondary pollen presentation genus-by-genus in 25 families. The subject has never been extensively reviewed, although secondary pollen presentation occurs in the largest family of flowering plants, the Asteraceae (Compositae), and a large family of great economic importance, the Leguminosae. Now material from the scattered literature is brought together and supplemented with original observations. Many species are illustrated and each family is individually discussed. The last two chapters provide an overview of the whole topic. All the main functions that secondary pollen presentation may perform can be carried out in other plants without it. It is concluded from this that the evolution of secondary pollen presentation has been subject to constraint and canalization. The floral biology of most plants with secondary pollen presentation has not been adequately studied. Appendix 1 points to a wide range of topics on which research at various technical and academic levels is needed. The book should also become a reference work for morphologists, systematists, and floral ecologists.

Springer-Verlag Wien New York

Sachsenplatz 4–6, P.O.Box 89, A-1201 Wien · 175 Fifth Avenue, New York, NY 10010, USA
Heidelberger Platz 3, D-14197 Berlin · 37-3, Hongo 3-chome, Bunkyo-ku, Tokyo 113, Japan

Plant Systematics and Evolution

Entwicklungsgeschichte und Systematik der Pflanzen

Continuation of Österreichische Botanische Zeitschrift

The intent of **Plant Systematics and Evolution** is to serve as a medium for world-wide scientific communication in the fields of plant morphology and systematics in the widest sense. On the basis of comparative, developmental, and functional approaches, contributions on the structure of lower and higher plants, from the electron-microscopic to the cytological, anatomical, and morphological level are encouraged. Great importance is attached to studies on the biology, ecology, distribution, systematics, phylogeny, and evolution of various groups. Modern methods of cytogenetics, molecular biology population analysis, electron microscopy, chemosystematics, palynology, cladistics, and numerical analysis are favoured. Also, emphasis is given to general aspects of evolutionary differentiation and mechanisms.

Subscription Information:
1994. Vols. 189-193 (4 issues each):
DM 2110,–, öS 14770,–, plus carriage charges
ISSN 0378-2697. Title No. 606

Sachsenplatz 4–6, P.O.Box 89, A-1201 Wien · 175 Fifth Avenue, New York, NY 10010, USA
Heidelberger Platz 3, D-14197 Berlin · 37-3, Hongo 3-chome, Bunkyo-ku, Tokyo 113, Japan

The manufacturer's authorised representative in the EU is Springer
Nature Customer Service Centre GmbH, Europaplatz 3, 69115 Heidelberg,
Germany. If you have any concerns regarding our products, please
contact ProductSafety@springernature.com

Printed and bound by CPI Group (UK) Ltd, Croydon, CR0 4YY

28/04/2026

02098462-0009